Introduction to Photonic Crystals

Introduction to Photonic Crystals

S. Sankararaman

ADHYAYAN

ADHYAYAN PUBLISHERS & DISTRIBUTORS
NEW DELHI (INDIA)

Published by
ADHYAYAN PUBLISHERS & DISTRIBUTORS
4378/4B, 105, J.M.D. House, Murari Lal Street
Ansari Road Darya Ganj, New Delhi - 110002
Ph.: 011-23263018, 011-23277156, Fax: 011-23280028
Email: adhyayanpublishers@yahoo.com

Introduction to Photonic Crystals

Edition 2011
ISBN 978-81-8435-252-8

Printed in India

Published by Harish Chandra Yadav for Adhyayan Publishers & Distributors, Laser Typeseting at A.V. Rao and Printed at Tarun Offset Printers, Delhi

Dedicated to
My Parents and Teachers

Acknowledgment

First of all I would like to thank *Satravedi* for publishing this book.

I express my gratitude to my students Ms. Jizo Joy and Manju Mathew who have helped me a lot in the preparation of this book. I am also thankful to Rev.Fr. Thomas Malekudy, the Manager and Fr. Paul Nedumpurath, the Principal, Newman College, Thodupuzha.

I also express my grateful thanks to Sri. Adv. P.T. Thomas, Sri. Adv. E.M.Agsthi, and *Sastravedi* members for their constant motivation in writing a science book.

S. Sankararaman

PREFACE

The term "**photonic crystals**" refers to a new class of artificial optical materials that will almost certainly form the basis of revolutionary new technologies in the 21st century. Photonic crystals are periodic optical (nano)structures that are designed to affect the motion of photons in a similar way that periodicity of a semiconductor crystal affects the motion of electrons. Photonic crystals occur in nature and in various forms have been studied by science for the last 100 years

A photonic band gap material is a dielectric material in which the dielectric constant is made to vary periodically. This periodicity produces a set of forbidden energies for photons in the material; the range of forbidden photon energies constitutes the so-called photonic band gap of the material. The photonic band gap is a frequency interval over which the linear electromagnetic propagation effects have literally been turned off. It is given this name because it is analogous to the electronic band gap in insulators and semiconductors. The term electronic band gap refers to the forbidden ranges of energy for electrons in these types of solids. So, the difference is that electronic band gap materials have bands of forbidden electron energies, while photonic band gap materials have bands of forbidden photon energies.

Since the invention of the laser, the field of photonics has progressed through the development of engineered materials which mold the flow of light. Photonic band gap (PBG) materials or Photonic crystals are a new class of dielectrics which are the photonic analogues of semiconductors. They represent a new frontier in quantum optics and offer many new technological application. Unlike semiconductors which facilitate the coherent propagation of electrons, PBG materials facilitate the coherent localization of photons.

The PBG exhibits a rich variety of nonlinear optical propagation phenomena. These include classical gap solitons and quantum gap solitons. These solitons may be important in the transmission of information through the otherwise impenetrable PBG. Now photonic crystals have started playing a significant role in the realm of science and technology.

Dr. Sajeev John's research papers are the real motivation behind this work. The book is written keeping the beginners in Photonics in mind. It is intended for creating an interest among science students in the latest field of Photonics. Since this is only an introductory book, mathematical intricacies have been avoided. Enough references given at the end of each chapter will be helpful for those who wish to know more about Photonic crystal. The book is prepared from various research articles published in journals. There are certain deliberate repetitions of some basic concepts as it will facilitate the learning process.

CONTENTS

CHAPTER 1

THE INVENTORS

Chapter contents

1.1 Introduction

The simplest form of a photonic crystal is a one-dimensional periodic structure, such as a multilayer film (a *Bragg mirror*). Electromagnetic wave propagation in such systems was first studied by **Lord Rayleigh** in 1887 [1], who showed that any such one-dimensional system has a bandgap.

A detailed theoretical study of one-dimensional periodic optical structures was performed by **V.P. Bykov [2].** Bykov investigated an effect of a photonic band gap on the spontaneous emission of atoms and molecules embedded inside a one-dimensional photonic crystal. Presumably, he was the first to suggest that the active medium of a laser may have a three-dimensional periodic structure.

Eli Yablonovitch and **Sajeev John** were the two physicists who invented the field of photonic crystals in 1987. In 1987 Eli Yablonovitch and Sajeev John [3,4] almost simultaneously published two milestone papers in Physical Review Letters.

The biographies of these eminent scientists are given below.

1.2 John William Strutt Rayleigh

John William Strutt Rayleigh (1842-1919), British mathematician, physicist, and Nobel laureate, is known for his research in wave phenomena. Rayleigh was born in Lanford Grove, near Maldon, England, and educated at the University of Cambridge. He served as professor of experimental physics and director of the Cavendish Laboratory at Cambridge from 1879 to 1884 and as professor of natural philosophy at the Royal Institution, London, from 1887 to 1905. He became Chancellor of Cambridge in 1908.

Rayleigh engaged in research into physical optics, light, color, and electricity and the dynamics of resonance and vibrations of gas and elastic solids. He was also responsible for the determination of electrical units of measurement. In 1894 he and the British chemist Sir William Ramsay discovered the inert element argon. For this discovery Rayleigh was awarded the 1904 Nobel Prize in physics. Ramsay won the Nobel Prize in chemistry the same year.

1.3 Eli Yablonovitch

Eli Yablonovitch along with Sajeev John was one of the two physicists who invented the field of photonic crystals in 1987. In addition to pioneering photonic crystals, he was the first to recognize that a strained quantum well laser has a significantly reduced threshold current copared to its unstrained counterpart. This technique is now aplied to the majority of semi-conductor lasers fabricated throughout the world.

He was a professor of electrical engineering at University of California, Los Angeles (UCLA) and continues to study and develop photonic crystals and photonic bandgap materials. As of July 2007, he has joined the Electrical Engineering and Computer Science department at UC Berkeley. His other researches includes silicon photonics, quantum computing, telecommunications, and surface plasmon optics.

He is also a member of both the National Academy of Sciences and the National Academy of Engineering.

He graduated with the Ph.D. degree in Applied Physics from Harvard University in 1972. He worked for two years at Bell Telephone Laboratories, and then became a professor of Applied Physics at Harvard. In 1979 he joined Exxon to do research on photovoltaic solar energy. Then in 1984, he joined Bell Communication Research, where he was a distinguished Member of Staff, and also Director of Solid-State Physics Research. In 1992 he joined the University of california, Los Angeles, where ge is now The Northorp Grumman Opto-Electronics Chair, Professor of Electical Engineering.

Prof. Yablonovitch's work has covered a broad variety of topics: non-linear optics, laser-plasma interaction, infrared laser chemistry, photovoltaic energy conversion, strained-quantum-well lasers, and chemical modification of semi conductor surfaces. Currently his main interests are in optoelectronics, high speed optical communications, high efficiency light-emitting diodes and nano-cavity lasers, photonic crystals at optical and microwave frequencies, quantum computing and quantum communication.

Yablonovitch was a Founder of the W/PECS series of Photonic Crystal International workshops that began in 1999. (PECS VII will be held in Australia in 2009.)

He is a Fellow of the Institute of Electrical and Electronic Engineers, the Optical Society of America, and the American Physical Society. Yablonovitch is a Life Member of Eta Kappa Nu, and a Member of the National Academy of Engineering and the National Academy of Sciences. He has awarded the Adolf Lomb Medal, the W. Streifer Scientific Achievement Award, the R.W. Wood Prize, and the Julius Springer Prize.

Research Interests

Professor Yablonovitch's research focuses on optoelectronics, high speed optical communications, nanocavity lasers, photonic crystals at optical and microwave frequencies, quantum computing and communication.

Awards and Recognitions

- 2005 Northrop Grumman Optoelectronics Chair in Electrical Engineering, UCLA
- 2004 Honorary Doctorate, Royal Institute of Technology, Sweden
- 2003-2004, Moore Distinguished Scholar, Caltech
- 2003 National Academy of Sciences
- 2003 National Academy of Engineering
- 2001 Julius Springer Prize in Applied Physics
- 1996 R.W. Wood Prize, Optical Society of America
- 1993 W. Streifer Achievement Award, IEEE Lasers & Electro-Optics Society
- 1992 Fellow, IEEE
- 1990 Fellow, American Physical Society
- 1990 Research and Development 100 Award
- 1982, Fellow, Optical Society of America
- 1978-79 Alfred P. Sloan Fellow
- 1978 Adolf Lomb Medal, Optical Society of America

Selected Publications

➢ E. Yablonovitch, M. Xiao, I. Martin, and H. Jiang, "Electrical Detection of the Spin Resonance of a Single Electron in a Silicon Field-Effect Transistor," *Nature*, vol. 430, no. 6998, pp. 435-39, July 2004.

➢ E. Yablonovitch, H. Jiang, H. Kosaka, H. Robinson, D. Rao, and T. Szkopek, "Optoelectronic Quantum Telecommunications Based on Spins in Semiconductors," *Proceedings of the IEEE*, vol. 91, no. 5, pp. 761-80, May 2003.

- E. Yablonovitch, H. Kosaka, D. Rao, H. Robinson, P. Bandaru, and K. Makita, "Single Photoelectron Trapping, Storage, and Detection in a Field Effect Transistor," *Physical Review B (Condensed Matter and Materials in Physics)*, vol. 67, no. 4, pp. 45104/1-5, Jan. 2003.
- E. Yablonovitch, "Photonic Crystals: Semiconductors of Light," *Scientific American*, vol. 285, no. 6, pp. 47-55, Dec. 2001.
- E. Yablonovitch, I. Gontijo, M. Boroditsky, S. Keller, U. Mishra, and S. DenBaars, "Coupling of InGaN Quantum-well Photoluminescence to Silver Surface Plasmons," *Physical Review B (Condensed Matter and Materials in Physics)*, vol. 60, no. 16, pp. 11564-67, Oct. 1999.

1.4 Sajeev John

Sajeev John is a University Professor at the University of Toronto and Government of Canada Research Chair holder. He received his Bachelors degree in physics in 1979 from the Massachusetts Institute of Technology and his Ph.D. in physics at Harvard University in 1984. His Ph.D. work at Harvard introduced the theory of classical wave localization and in particular the localization of light in three-dimensional strongly scattering dielectrics. From 1984–1986 he was a Natural Sciences and Engineering Research Council of Canada post doctoral fellow at the University of Pennsylvania as well as a laboratory consultant to the Corporate Research Science Laboratories of Exxon Research and Engineering from 1985-1989.From 1986-1989 he was an assistant professor of physics at Princeton University.

While at Princeton, he co-invented (1987), along with Eli Yablonovitch, the concept of photonic band gap materials, providing a systematic route to his original conception (1984) of the localization of light. He was a laboratory consultant to Bell Communications Research (Red Bank, NJ) in 1989. In the fall of 1989 he joined the senior physics faculty at the University of Toronto. He has been a Principal Investigator for Photonics Research Ontario, a Canadian center of excellence and is a Fellow of the Canadian Institute for Advanced Research. Professor John is the winner of the 2001 King Faisal International Prize in Science, which he shared with C.N. Yang. He is also the first ever winner of Canada's Platinum Medal for Science and Medicine in 2002. Dr. John is the winner of the Institute of Electrical and Electronics Engineers (IEEE) LEOS International Quantum Electronics Award in 2007 for "the invention and development of light-trapping crystals and elucidation of their properties and applications." Dr. John also received the 1996 Herzberg Medal for Physics and the 2007 Brockhouse Medal for Condensed Matter and Materials Physics from the Canadian Association of Physicists. He received the first ever McLean Fellowship from the University of Toronto in 1996, the 1997 Steacie Prize in Science and Engineering from the National Research Council of Canada, and the 2004 Rutherford Medal from the Royal Society of Canada. He is the first ever winner of Brockhouse Canada Prize in 2004, which he shared with materials chemist Geoff Ozin for their groundbreaking interdisciplinary work on photonic band gap materials synthesis. Professor John has also received the Killam Fellowship of the Canada Council for the Arts, the Guggenheim Fellowship (USA), the Japan Society for the Promotion of Science Fellowship, and the Humboldt Senior Scientist Award (Germany). In 2007, Dr. John was awarded the C.V. Raman Chair Professorship of the Government of India. Prof. John is a Fellow of the

American Physical Society, the Optical Society of America, the Royal Society of Canada, and a member of the Max-Planck Society of Germany.

Research Interests

Condensed matter physics, disordered systems, optical properties, localization, many-body theory, magnetism, superconductivity.

Professor John is one of the world's leading condensed-matter theorists. His work on the propagation of light in matter has transformed our understanding of that field. His name is synonymous with the field that has become known as "photonic crystals." The basic concept was developed in a series of papers by Professor John in the 1980s.

1.5 Bykov Vladimir Pavlovich (V.P. Bykov)

- Doctor of physics and mathematics
- Professor
- Leading scientist of General Physics Institute of the Russian Academy of Sciences (GPI RAS) Russia, 117588 Moscow.

Bykov V.P. was born in Moscow on 20.07.1931. Graduated from Moscow State University (Physical Faculty) in 1954 (honors diploma). He had a student practice in the Institute for Physical Problems (IPP) of Academy of Sciences of USSR - 1952-54. His Diploma thesis was on "Raman scattering of light by superconductors". 1954-1963 worked in the IPP in the laboratory of acad. In cooperation with S.P. Kapitza and V.N. Melekhin microtron with record current was created. In 1962 the degree of Candidate of physics and mathematics was conferred on by United Institute for Nuclear Research (Dubna). His Candidate thesis was " Investigation of the electron wavepackets of microtron."

In 1963-1976 worked in Laser Institute - "Polus". In this period the geometrical optics of laser resonators was developed and at first time spontaneous emission radiation in periodical structures was investigated. In 1975 the degree of Doctor of Physics and Mathematics was conferred on by Moscow Institute of Physics and Technology (PhysTech, Dolgoprudny, Moscow region). Doctoral thesis was "Eigenwaves and radiation in open systems." In 1976-1983 worked in Optical Laboratory (Director - acad. I.V. Obreimov) of the Institute of Nonorganic and General Chemistry of Ac Sci. USSR. There he worked on Field - quantum state of laser radiation, development of Scully-Lamb model, appearance of coherence in laser radiation. In 1983-till now is working in General Physics Institute (director acad. A.M. Prokhorov) of Russ. Acad. Sci., in lab. of acad. A.M. Prokhorov. His fields of interest include quantum properties of laser light, propagation of electromagnetic waves in layers and fibers, nature of photocounts, radiation detectors, QED. Since 1968 Bykov V.P. works in Moscow Institute of Physics and Technology (PhysTech, Dolgoprudny, Moscow region) as a teacher, after 1986 - as a professor (the professor degree was conferred on in 1986 by Moscow Institute of Physics and Technology). Now Bykov V.P. has more than one hundred publications. Between them four books - two in English and two in Russian.

Books:

1. Bykov V.P., "*Radiation of Atoms in a Resonant Environment,*" 1993, World Publishing, Singapore, Engl.
2. Bykov V.P., Silichev O.O., "*Laser Resonators*" Cambridge International Science Publishing, 1996, Engl.
3. Bykov V.P., Shepelev G.V. ,"*Radiation of atoms near macroscopic bodies,*" Moscow, Nauka, 1986, Russ.

4. Bykov V.P., *"Quantum properties of laser radiation ,"* PhysTech Publishing 1986, Moscow, Russ.

1.6 History of Localization and Photonic bandgap

1918: Lord Rayleigh suggested that the iridescence of some insects arose from interference effects

1958: Electron Localization: All electronic states localize P.M. Anderson, *Phys.Rev.* 109,1492 (1958)"Absence of Difffusion in certain Random Lattices."

1983: Photon Localisation : Some states are Localized S. John and M.J. Stephen, *Phys.Rev.*, B28,6358 (1983)

1984: Photon Localisation, S. John, *Phys.Rev. Lett.* 53,2169, (1984)

1985: Photon Localisation P.W.Anderson, *Phil. Mag.* B52,505 (1985) Weak Localization of light (Coherent backscattering), M.P. Von Albada and A. Lagendjik *Phys.Rev.Lett.* 55,2692 (1985)

P.E. Wof and G. Maret, *Phys.Rev.Lett.*, 55,2696 (1985)

1987: *Prediction of Photonic bandgap (PBG)* by S. John and E. Yablonovitch

1990: Computational Demonstration of PBG, K.M. Ho, C.T. Chan, C.M. Soukoulis, *Phys.Rev.Lett.* 65,3152 (1990)

Quantum Electrodynamics in PBG,S.John and J. Wang, *Phys.Rev.Lett.* 74,2418 (1990)

1991: Experimental Demonstration of Microwave Localization and PBG 2D system: S. Shultz et,al., *Nature* 354,53 (1991)

3D system: AZ. Genack and N. Garcia, *Phys.Rev. Lett.* 67,2295 (1991)

1994: Experimental observation of "Laser Paint", N.M. Lawandy et,al., *Nature* 368,436 (1994)

1995: "Large" Scale 2D PBG Macroporous Silicon U. Gruning,V. Lehman, C.M. Englehardt ,*Appl. Phys. Lett.*, 66,3254 (1995)

1995: Present: "Woodpile structures" E.O. Zbay,S. Noda, S. Lin

1996: U. Gruning et al., *Appl.Phys.Lett.*, 68,3254 (1996)

1997: Experimental Demonstration of Light Localization, D. Wiersma etal., *Nature*, 360, 671 (1997).

1998: Present: Inverted Opals, TiO_2, CdSe, Ge, Si, GaP ....

2000: Parker, A.R., "515 million years of structural color." *J. Optics A: Pure and Applied Optics* (2000) **2**, R15-28

2001: Square spirals SiO_2, Si

2004: SEM investigations on Butterfly Balint et,al., *Issues on Multidisciplinary Microscopy Research and Education*, 87 (2004)

2005: Photonic crystal coat in Beetles, Vigneron et al., *Physical Review* E, December 2005

1.7 References

1. J.W.S. Rayleigh, "On the remarkable phenomenon of crystalline reflexion described by Prof. Stokes." *Phil. Mag.*, 26, 256-265. (1888)
2. V.P. Bykov,"Spontaneous emission from a medium with a band spectrum." *Sov. J. Quant.Electron.*,4, 861-871 (1975)
3. The research group of Prof. W.L. Vos in the Netherlands P. Lodahl, A.F. van Driel, I.S. Nikolaev, A. Irman, K. Overgaag, D.Vanmaekelbergh, and W.L. Vos "Controlling the dynamics of spontaneous emission from quantum dots by photonic crystals" *Nature* 430, 654-57 (2004) http cops.tnw.utwente.nl/ pdf/04/nature 02772. pdf

4. Review: S. Johnson (MIT) Lecture 3: Fabrication technologies for 3d photonic crystals, a survey http://ab-initio.mit.edu/photons/tutorial/L3-fab.pdf
5. Encyclopedias – Britannica and Encarta

CHAPTER 2

PHOTONIC BANDGAP MATERIALS

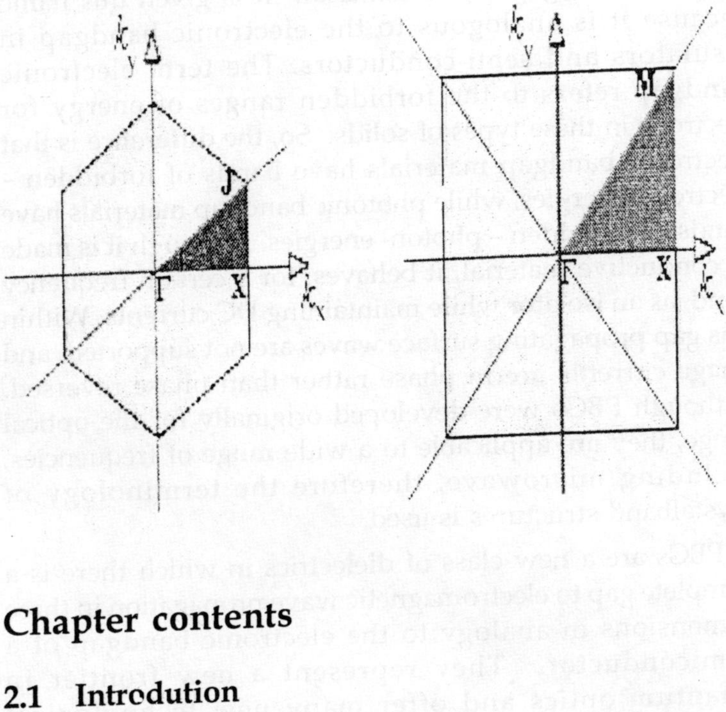

Chapter contents

2.1 Introduction

A photonic bandgap material (PBG) is a dielectric material in which the dielectric constant is made to vary periodically. This periodicity produces a set of forbidden energies for photons in the material; the ranges of forbidden photon energies constitute the so-called photonic bandgap of the material. It is given this name because it is analogous to the electronic bandgap in insulators and semi-conductors. The term electronic bandgap refers to the forbidden ranges of energy for electrons in these types of solids. So, the difference is that electronic bandgap materials have bands of forbidden - electron - energies, while photonic bandgap materials have bands of forbidden – photon- energies. Although it is made of conductive material, it behaves, for a certain frequency band, as an isolator while maintaining DC currents. Within this gap propagating surface waves are not supported, and image currents are in phase rather than phase reversed. Although PBGs were developed originally for the optical range, they are applicable to a wide range of frequencies, including microwave, therefore the terminology of crystalband structures is used.

PBGs are a new class of dielectrics in which there is a complete gap to electromagnetic wave propagation in three dimensions in analogy to the electronic bandgap of a semiconductor. They represent a new frontier in quantum optics and offer many new technological applications. A number of new physical phenomena have been predicted to occur in these materials including strong localization of light; inhibited spontaneous emission of atoms, and photon-atom bound states. A three-dimensional PBG material is a periodic dielectric system with an absolute frequency gap for electromagnetic waves. The opening of photonic gaps is a delicate balance between both refraction index contrast and topology of the underlying lattice. PBG materials have attracted much attention due to their many remarkable applications, especially for gaps in the visible range.

A perfect isolation of electromagnetic modes is possible if a localized state of light can be formed. It has been suggested that such states may arise in strongly scattering dielectric microstructures. In a periodic, dielectric array, this can be facilitated in a synergetic interplay between micro cavity resonances of individual dielectric particles and the Bragg scattering resonance of the array. This leads to the formation of a PBG. It was suggested that this would be accompanied by the inhibition of spontaneous emission as well as existence of photonic band gaps has been predicted that the excited atom forms a photon-atom bound state, the optical analog of an electron-impurity level bound state in the gap of a semi-conductor.

PBG materials forbid propagation of photons within a certain frequency range irrespective of their direction of propagation, permitting the possibility of studying new physics within the gap. For example, spontaneous emission and radiative electron hole recombination will be strongly inhibited if the frequency of the photon to be emitted lies in the forbidden gap. The concept of a three-dimensional photonic band structure was introduced by Yablonovitch as an extension of a familiar one-dimensional optical device: a Fabry–Perot resonator constructed from dielectric mirrors.

2.2 Photonic Bandgap Structures

In 1987, Yablonovitch demonstrated the existence of PBGs, analogous to those well known in semi-conductor technology [1]. In a semi-conductor, free electrons are subject to periodic electric fields due to the arrangement of ions in the material, while in PBG structures photons are subject to a periodic dielectric constant [2]. The key factor in these materials is that a unit cell in the periodic arrangement must be on the order of the wavelength. With this in mind, Yablonovitch first showed these bandgaps using microwaves since this allowed the structure to be fabricated on a larger scale [3]. This discovery, which proved that periodic microstructures in a dielectric medium can create PBG, has led researchers to pursue a multitude of applications ranging from integrated optical circuits, to a new kind of optical fiber.

The presence of bandgaps in an optical material essentially means the same in photonics as it does in electronics, regions or sets of forbidden energies. Since no solutions for the electromagnetic field exists in these regions, a microstructered material designed in a particular manner can guide light in ways never before realized with traditional wave-guide technology. For applications of integrated optical circuits, these new materials promise the ability to, among other things, guide light around sharp bends. This effect has recently been demonstrated using millimeter wavelengths to make a lossless transition around a 90-degree turn [4]. While PBG materials are set to make remarkable advances in many optical devices, it will be the use of these materials as fiber optic delivery systems. The use of photonic crystals as wave-guides is a fairly new study [1-7], which may eventually outperform traditional fiber optics. So far, microstructures optical fibers have been studied in two different forms; one, which is similar to standard total internal reflection guidance, and another form, which guides light based on the presence of PBGs.

2.3 The Origin of the Photonic Bandgap

A complete photonic bandgap is a range of ω in which there are no propagating (real vector k) solutions of Maxwell's equations for any k, surrounded by propagating states above and below the gap. There are also incomplete gaps, which only exist over a subset of all possible wavevectors, polarizations, and/or symmetries. Both sorts of gaps are discussed in the subsequent sections, but in either case their origins are the same, and can be understood by examining the consequences of periodicity for a simple one-dimensional system.

Consider a one-dimensional system with uniform $\varepsilon = 1$, which has Plane wave eigen solutions ω (k) = $c\kappa$, as depicted in Figure 2.1 (left). This ε has trivial periodicity a for any a $\geq$ 0, with a = 0 giving the usual unbounded dispersion relation. We are free, however, to label the states in terms of Bloch envelope functions and wavevectors for some a $\neq$ 0, in which case the bands for $|k| > \pi/a$ are translated ("folded") into the first Brillouin zone, as shown by the dashed lines in Figure 2.2 (left). In particular, the k = $-\pi/a$ mode in this description now lies at an equivalent wavevector to the k $= \pi/a$ mode, and at the same frequency; this accidental degeneracy is an artifact of the "artificial" period we have chosen. Instead of writing these wave solutions with electric fieldsm $E(x) \sim \exp(\pm i\pi x/a)$, we can equivalently write linear combinations $e(x) = \cos(\pi x/a)$ and $o(x) = \sin(\pi x/a)$ as shown in Figure. (2.3), both at ω = $c\pi/a$.

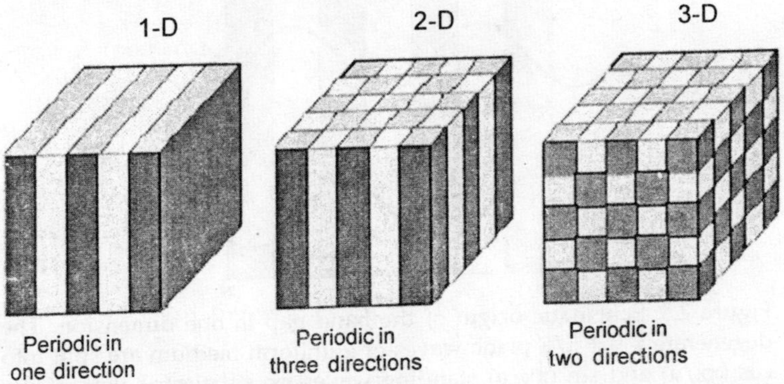

Figure 2.1: Schematic depiction of photonic crystals periodic in one, two, and three directions, where the periodicity is in the material (typically dielectric) structure of the crystal. Only a 3-D periodicity, with a more

complex topology than is shown here, can support an omni-directional photonic bandgap.

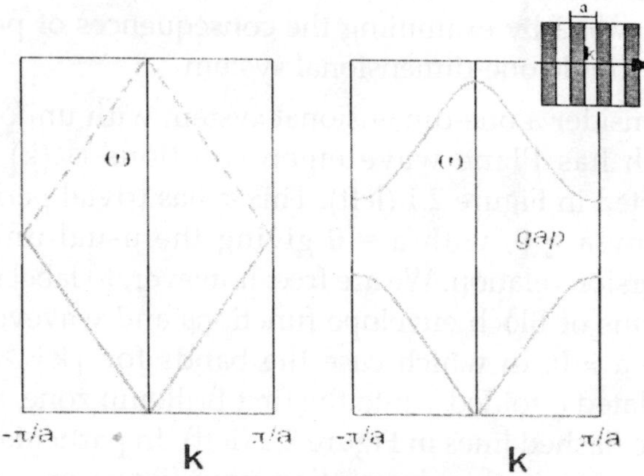

Figure 2.2: Left: Dispersion relation (band diagram), frequency ω versus wave number k, of a uniform one-dimensional medium, where the dashed lines show the "folding" effect of applying Bloch's theorem with an artificial periodicity a. Right: Schematic effect on the bands of a physical periodic dielectric variation (inset), where a gap has been opened by splitting the degeneracy at the $k = \pm\pi/a$ Brillouin-zone boundaries (as well as a higher-order gap at $k = 0$).

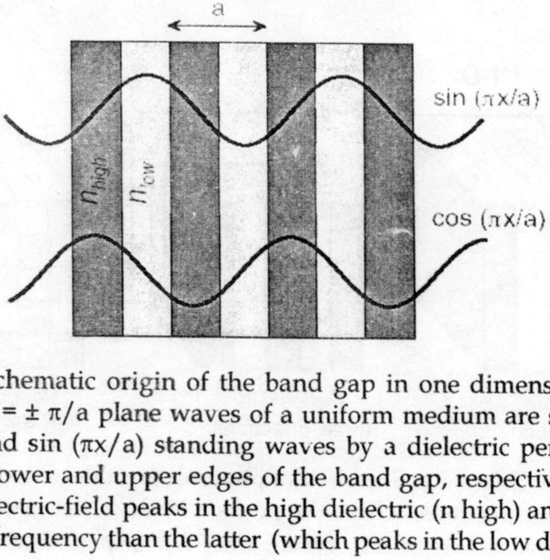

Figure 2.3: Schematic origin of the band gap in one dimension. The degenerate $k = \pm\pi/a$ plane waves of a uniform medium are split into cos (πx/a) and sin (πx/a) standing waves by a dielectric periodicity, forming the lower and upper edges of the band gap, respectively—the former has electric-field peaks in the high dielectric (n high) and so will lie at a lower frequency than the latter (which peaks in the low dielectric).

Now, however, suppose that we perturb ε so that it is nontrivially periodic with period a; for example, a sinusoid $\varepsilon(x) = 1+\Delta \cos(2\pi x/a)$, or a square wave as in the inset of Figure (2.2). In the presence of such an oscillating "potential," the accidental degeneracy between ε (x) and o(x) is broken: supposing $\Delta > 0$, then the field ε (x) is more concentrated in the higher - ε regions than o(x), and so lies at a lower frequency. This opposite shifting of the bands creates a band gap, as depicted in Figure 2.2 (right). (In fact, from the perturbation theory described subsequently, one can show that for $\Delta \ll 1$ the band gap, as a fraction of mid-gap frequency, is $\Delta\omega/\omega \approx \Delta/2$.)

It has long been known by diamantists, and used to check for authenticity, that dipping a gem in a solution with a matching index of refraction turns it invisible. This is due to the fact that what makes a homogeneous object's boundaries visible is the reflection and refraction of light therein. Therefore if light does not find difference in traversing a surface it will not be in any way scattered, as it is not in a homogeneous medium. The main feature of PXs is the periodic modulation of such property (dielectric constant) along one, two or three directions of space (see Figure.2.4). In a composite formed by two dielectrics, we will consider the scattering center that in which light propagates more slowly, i.e. that of higher ε. If the scattering centers are regularly arranged in a medium, light is coherently scattered. In this case, interference will eventually cause that some frequencies will not be allowed to propagate, giving rise to forbidden and allowed bands. Under certain conditions that will be detailed later in this section, regions of frequency may appear that are forbidden regardless of the propagation direction in the PX (see Figure 2.4). In such case, this PX is said to present a full PBG. On the contrary, if the forbidden photonic band varies with the propagation direction in the PX, a photonic pseudo gap is spoken of. What is more, by introducing defects in the PX, we can introduce allowed energy levels in the gap, as occurs when a semiconductor is doped. All these facts permit to establish a parallelism between the formalism used for electrons in ordinary crystals and that for photons in PX.

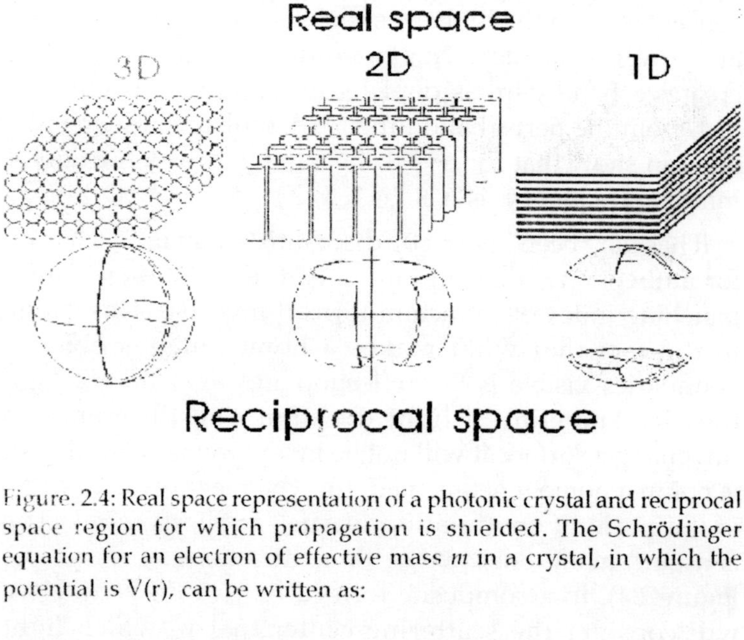

Figure. 2.4: Real space representation of a photonic crystal and reciprocal space region for which propagation is shielded. The Schrödinger equation for an electron of effective mass m in a crystal, in which the potential is V(r), can be written as:

$$\left[\frac{-\hbar^2}{2m}\nabla^2 + V(r)\right]\psi(r) + E\psi(r) \quad (2.1)$$

where V(r) is a periodic function with the periodicity of the lattice, R:

$$V(r) = V(r+R) \quad (2.2)$$

The eigenstates of this equation are also periodic functions with period R. The dispersion relationship derived, E (k), will present a forbidden band for all energies E which have imaginary values. Similarly, in a medium in which a spatial modulation of the dielectric constant e (r) exists, photon propagation is governed by the classical wave equation for the magnetic field H(r):

$$\nabla X\left[\frac{1}{\varepsilon(r)}\nabla xH(r)\right]=\left[\frac{\omega}{c}\right]^2 H(r) \qquad (2.3)$$

In a PX, ε (r) is a periodic function:

$$\varepsilon\,(r) = \varepsilon(r+R) \qquad (2.4)$$

These equations show the parallelism between electrons in crystalline solids and photons in PX. In Figure. (2.5) it is shown how gaps are developed in both an electronic crystal and a PBG material.

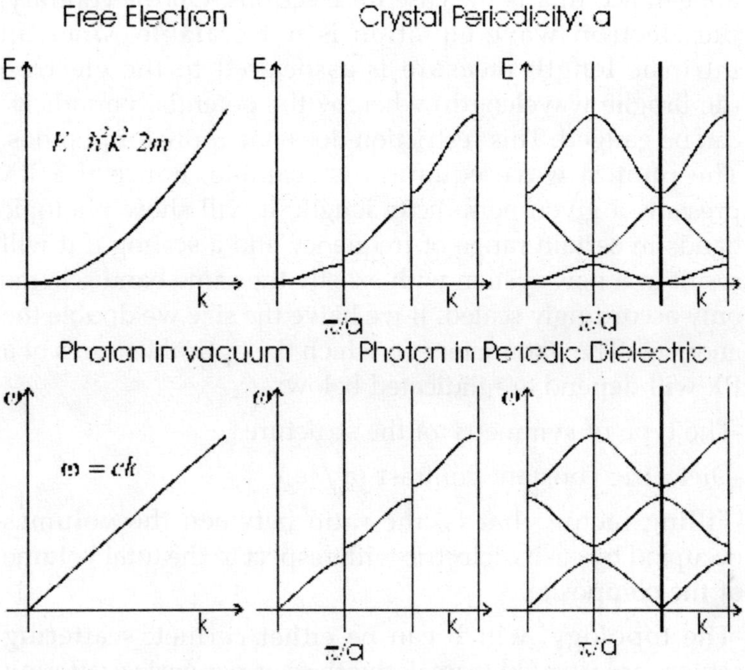

Figure. 2.5. Energy dispersion relations for free electron and electron in a (1D) solid and for a free photon and a photon in a PX.

The energy dispersion relation for an electron in vacuum is parabolic with no gaps. When a periodic potential is present gaps open and electrons with energies therein have localized (non-propagating) wave functions as opposed to those of electrons in allowed bands, which have extended (propagating) wave functions. In a similar way, a periodic dielectric medium will present frequency regions where propagating photons are not allowed and will find it impossible to travel the crystal. One important difference between electrons and photons rests on the different nature of their associated waves. Electrons are scalar waves, while photons are vectorial ones. This implies that, in the latter case, polarization must be taken into account. This finally results in much more restrictive conditions for gap appearance that is the case for electrons. On the contrary, the electron wave equation is not scalable, since an intrinsic length measure is associated to the electron (de Broglie wavelength) whereby the potential periodicity can be gauged. This restriction does not apply for photons. The photon wave equation is scalable, hence if a PX presents a given periodicity length, it will show photonic bands in certain range of frequency and a scaling if it will result in a new system with exactly the same band scheme only accordingly scaled. If we halve the size we double the energies. The parameters on which the optical features of a PX will depend are indicated below:

-The type of symmetry of the structure.

-Dielectric constant contrast ($\varepsilon_1/\ \varepsilon_2$).

-Filling factor, that is, the ratio between the volumes occupied by each dielectric with respect to the total volume of the composite.

-The topology, which can be either cermet: scattering centers are isolated from each other; or network: scattering centers are connected between them.

-The shape of the scattering centers.

All these factors determine the photonic band structure of the PX and, therefore, its optical properties. Economou and Sigalas have published a general discussion about topologies in PBG theory and the generalization to other classical waves. Recently, it has been shown that a slight modification of either the symmetry or the shape of the scatterer can enlarge the gap.

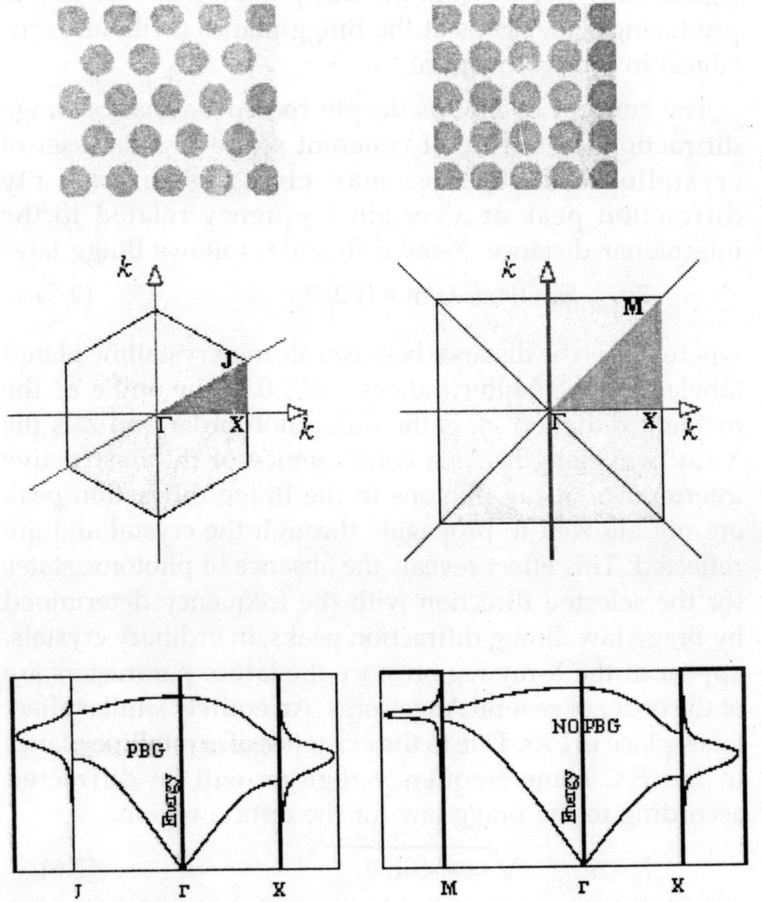

Figure. 2.6: Real and reciprocal space representation of triangular and square two-dimensional crystals to highlight the formation of a full bandgap and a pseudogap.

In Figure 2.6 it can be seen that high symmetry favors the photonic gap appearance. In this case the higher symmetry of the hexagonal lattice is reflected by the fact that the Brillouin zone is closer to circular than the square lattice: distances in reciprocal space to the principal points of the irreducible zone are similar. It is obvious that a high dielectric constant contrast produces the optical bands significantly depart from the free photon behavior (ω=ck) producing wider gaps at the Bragg planes (Brillouin zone edges) in reciprocal space.

The concept of PBG is deeply rooted on that of Bragg diffraction. By virtue of coherent scattering each set of crystallographic planes may give rise to an X-ray diffraction peak at a certain frequency related to the interplanar distance. X-ray diffraction follows Bragg law:

$$2d_{(hkl)} \text{ Sin } \theta = m\lambda;\ \text{m} = 1, 2, 3, \qquad (2.5)$$

where $d_{(hkl)}$ is the distance between atomic crystalline planes labeled by the Miller indices *(hkl)*, θ is the angle of the incident radiation, m is the diffraction order and λ is the X-ray wavelength. As a consequence of the destructive interference, X-ray photons in the Bragg diffraction peak are not allowed to propagate through the crystal and are reflected. This effect reveals the absence of photonic states for the selected direction with the frequency determined by Bragg law. Bragg diffraction peaks, in ordinary crystals, appear at the X-ray region since the lattice parameters are of the order of several Angstroms. An entirely similar effect takes place in PXs. Due to the existence of crystalline planes in the PX, some frequency regions will be diffracted according to the Bragg law for the optical region:

$$\lambda_c = 2d_{(hkl)} \sqrt{<\varepsilon> - \text{Sin}^2\theta_{(hkl)}} \qquad (2.6)$$

where λ_c is the wavelength of the EM wave, $d_{(hkl)}$ the interplanar distance for the *(hkl)* crystallographic direction, $<\varepsilon>$ the average dielectric constant of the PX and $\theta_{(hkl)}$ the angle between the incident radiation and the normal to the set of crystalline planes determined by the *(hkl)* indices. An important difference between X-ray diffraction in solids and diffraction in PXs is the bandwidth of the Bragg peaks. X-ray diffraction peaks are extremely narrow ($\Delta\lambda/\lambda \approx 10^{-6}$) and mainly instrumentally broadened. In PXs, the diffraction condition for a given direction of the wave vector is met for a larger range of frequencies ($\Delta\lambda/\lambda \approx 10^{-2}$). This originates mainly in the different contrast of dielectric constants occurring for these two well apart wavelengths: indices of refraction for X-ray radiation scarcely differ from unity whereas at optical wavelengths they are rather larger. Eventually, Bragg peaks in PXs become so broad that can overlap others coming from different crystallographic planes. This is schematically depicted in Figure 2.6 where we show that this occurs in a hexagonal symmetry as opposed to the square symmetry in which the gaps do not overlap. Consequently, it could be possible to find a certain frequency region where the photon is not allowed to propagate regardless of direction. A material with these properties is called a PX with a full bandgap.

2.4 Photonic Bandgap Formation in Certain Self-Organizing Systems

Photonic band structure in certain self-organizing systems that self-assemble into large-scale photonic crystals with PBGs or pseudogaps in the near-visible frequency regime is discussed here. These include colloidal suspensions, inverted opals, and acroporous silicon. It is found that complete three-dimensional PBGs spanning roughly 10 percent and 15 percent of the gap center frequency are attainable by incomplete infiltration of an opal with silicon and germanium, respectively. The PBG band structure of both face center cubic and hexagonal close packed photonic crystals gets modified by sintering the opal prior to infiltration and by applying strain along various crystallographic directions. When the total photon densities of states as well as the local density of states (LDOS) are evaluated within the photonic crystal it is found that the LDOS may exhibit considerable pseudogap structure even for systems that do not exhibit a complete PBG. These results are directly relevant to quantum optical experiments in which atoms, dye molecules, or other active materials are inserted into specific locations within the photonic crystal. When the resonant optical transition of these dopants is tuned close to a pseudogap or other abrupt structure in the LDOS, novel effects in radiative dynamics associated with a "colored vacuum" may be realized. PBG materials facilitate the coherent localization of light [8-10], leading to novel phenomena in quantum optics as well as important technological applications.

Although the band structure of classical scalar waves readily yields complete three-dimensional gaps for simple structures such as a face centered cubic lattice of spherical scatterers [11], the vector nature of the electromagnetic wave equation leads to much more restrictive conditions on the dielectric microstructure for the formation of a PBG. Initial difficulties in designing structures that exhibit a complete PBG were allayed by the discovery that the diamond lattice of dielectric spheres exhibits such a frequency range for which (linear) wave propagation is forbidden [12]. In the microwave regime, other diamond likestructures obtained by drilling cylindrical holes in a bulk dielectric material (with a refractive index of 3.5) havebeen demonstrated to exhibit band gap to center frquency ratios $D\omega/\omega_0$ as large as 20 percent [13]. Since then, numerous structures amenable to layer-by-layer fabrication and drilling that possess complete PBG have been suggested [14-17]. A number of structures have already been fabricated with PBGs in the range of millimeter waves [18-20].

These structures not only confirm the soundness of the concept of a PBG but also have interesting applications in the microwave to millimeter wave range. For instance, an antenna mounted on a conventional dielectric constant substrate radiates the majority of its radiation into the substrate itself. If the substrate is engineered into the form of a PBG material with a gap at the radiation frequency, the losses can be minimized, leading to highly directional transmitters [21, 22]. Other applications include angular filters [23] and cladding material for preventing losses in waveguide structures that contains bends or junctions [24-26]. Nevertheless, it is for visible and near-infrared frequencies where PBG materials are likely to have their most important impact. For example, applications in telecommunications may require the fabrication of large-scale three-dimensional PBG materials with a gap centered around 1.5 mm. These applications may include the design of zero threshold microlasers, light emitting diodes that exhibit coherence properties at the single photon level, subpicosecond optical switches, and all-optical transistors. In addition, PBG materials represent a frontier in photon-atom interaction phenomena and non-linear optics. The single greatest obstacle to realizing the potential of PBG materials has been the lack of an inexpensive and reliable means of microfabricating large-scale three-dimensional materials with sizeble gaps at near visible frequencies. Here the possibility of overcoming this obstacle using self-assembling three-dimensional structures based on "inverted opals" is discussed. Starting from a close-packed face centered cubic lattice of SiO_2 spheres (opal) with diameter on the scale of a micrometer, PBG materials with gaps in the range 5-15 percent of the center frequency may be realized by infiltration of the opal with high refractive index materials such as Si or Ge and subsequently removing the SiO_2 by chemical etching. These structures may exhibit a number of novel effects in quantum and non-linear optics.

While linear wave propagation is absent in the gap of a PBG material, non-linear propagation effects in the form of ultrashort solitary wave pulses can still occur. Recent studies indicate that as a result of their large group velocity dispersion near a photonic band edge and complex symmetries, PBG materials exhibit a much richer variety of non-linear wave propagation phenomena than conventional waveguides or optical fibers [27-29]. In addition, PBG materials exhibit different quantum optical features, related to the drastic alteration of the photon density of states (DOS). A vanishing DOS leads to bound photon-atom states [30, 31], suppressed spontaneous emission [32, 33], and strong localization of photons [34]. Unlike conventional microcavities and other confined optical systems, PBG materials support propagation effects on the scale of the localization length for photons leading to different types of cooperative behavior involving photons and atoms. In a PBG material, photon localization effects render the master equation approach of standard quantum optics problematic since localization implies highly non-Markovian memory effects in radiative dynamics. In addition, it requires a non-standard treatment that goes beyond the Born approximation [35]. Among the phenomena predicted in this frontier of quantum optics are (i) collective switching of two-level atoms from the ground to the excited state with low intensity applied laser fields leading to all-optical transistor action (ii) anomalous super radiant emission rates and low-threshold band-edge lasers (iii) low-threshold non-linear optical response, and (iv) highly non-classical states of light within the PBG in the form of multiphoton localization and propagating quantum gap solitons [36].

These very exciting prospects for PBG materials with band gaps in the visible region have triggered numerous efforts to manufacture such structures, with periodicity on the optical scale. Two-dimensional PBG materials [37-40] have reached bandgaps (complete in two dimensions and incomplete in three dimensions) around wavelengths of 5ìm and the (electrochemical etching) technique is clearly capable of further downscaling of the structures. In three dimensions a number of self-assembling photonic crystals already exist. These include colloidal systems and artificial opals. Unfortunately, these readily available photonic crystals do not satisfy the necessary criteria of high index contrast and correct network topology to produce a complete PBG. Theoretical studies, however, indicate the possibility of complete PBG in closely related structures. Binary colloidal mixtures consisting of two sizes of TiO_2 spheres as well as fcc lattices consisting of low dielectric inclusions in a connected high dielectric network (henceforth called inverted structures) exhibit sizeble PBGs. The latter possibility has spurred highly interesting work on emulsion templating and on the infiltration of opals with silicon or germanium. These techniques will soon lead to a large-scale PBG material with a complete three-dimensional gap at near-visible frequencies.

2.5 References

1 Barkou, Broeng, and Bjarklev, *Guidance of Light Alongan Air Column in a New Class of Optical Fibers*. DOPS-NYT 1-1999

2. Jang, Study of Photonic Crystal Based Waveguide and Channel Drop Filter and Localization of Light in Phonic Crystal. Masters thesis, University of Texas, Austin. May 2000.

3. Kell yand Sheerin, Photonic Crystals www.physics. dcu.ie/~jpm/PS407/photonic.pdf.

4. Broeng et., l Waveguidance by the Photonic Bandgap Effect in Optical Fibers. *J. Opt. A. Pure Appl. Opt.* 1 (1999) (477-482)

5. Kuhlmey, Microstructured Optical Fibers: where's the edge? *Optics Express*, V. 10 No.22 Nov 2002.

6. Mortensen and Folkenberg, Modal Cutoff and the V Parameter in Photonic Crystal Fibers. *Optics Letters*, V. 28, No. 20 Oct. 2003.

7. Knight, Photonic Crystal Fibers, *Nature*, V. 424, August 2003.

8. S. John, *Phys. Rev. Lett.* 53, 2169 (1984)

9. S. John, *Phys. Today* 44 (5), 32(1991).

10. D. Wiersma, P. Bartelini, A. Lagendijk, and R. Righini, *Nature* (London) 390, 671 (1997).

11 S. John and R. Rangarajan, *Phys. Rev.* B 38, 10 101 (1988).

12. K.-M. Ho, C.T. Chan, and C.M. Soukoulis, *Phys. Rev. Lett.* 65, 3152 (1990).

13. E. Yablonovitch, T. J. Gmitter, and K. M. Leung, *Phys. Rev. Lett.* 67, 2295 (1991).

14. K.-M. Ho, C.T. Chan, C.M. Soukoulis, R. Biswas, and M. Sigalas, Solid State Commun. 89, 413 (1993).

15. C.T. Chan, S. Datta, K.-M. Ho, and C.M. Soukoulis, *Phys. Rev.* B 50, 1988 (1994).

16. S. Fan, P.R. Villeneuve, R. D. Meade, and J.D. Joannopoulos, *Appl. Phys. Lett.* 65, 1466 (1994).

17. K.M. Leung, *Phys. Rev.* B 56, 3517 (1997).

18. E. Ozbay, A. Abeyta, G. Tuttle, M. Tringides, R. Biswas, C.T. Chan, C.M. Soukoulis, and K.M. Ho, *Phys. Rev.* B 50, 1945 (1994).

19. E. Ozbay, E. Michel, G. Tuttle, R. Biswas, M. Sigalas, and K.M. Ho, Appl. Phys. Lett. 64, 2059 (1994).

20. E. Ozbay, G. Tuttle, J.S. McCalmont, R. Biswas, C.M. Souk-oulis, and K.M. Ho, *Appl. Phys. Lett.* 67, 1969 1995).

21. E.R. Brown, C.D. Parker, and O.B. McMahon, *Appl. Phys.Lett.* 64, 3345 (1994).

22. S.D. Cheung, R. Biswas, E. Ozbay, J.S. McCalmont, G. Tuttle, and K.M. Ho, *Appl. Phys. Lett.* 67, 3399 (1995).

23. M. Sigalas, J.S. McCalmont, K.M. Ho, and G. Tuttle, *Appl. Phys. Lett.* 68, 3525 (1996).

24. A. Mekis, J.C. Chen, I. Kurland, S. Fan, P. Villeneuve, and J.D. Joannopoulos, *Phys. Rev. Lett.* 77, 3787 (1996).

25. S.-Y. Lin, V.M. Hietala, S.K. Lyo, and A. Zaslavsky, *Appl. Phys. Lett.* 68, 3233 (1996).

26. K. Sakoda, T. Ueta, and K. Ohtaka, *Phys. Rev.* B 56, 14 905 (1997)

27. S. John and N. Akozbek, *Phys. Rev. Lett.* 71, 1168 (1993).

28. N. Akozbek and S. John, *Phys. Rev.* E 57, 2287 (1998).

29. N. Akozbek and S. John, *Phys. Rev.* E 58, 3876 (1998)

30. S. John and J. Wang, *Phys. Rev. Lett.* 64, 2418 (1990).

31. S. John and J. Wang, *Phys. Rev.* B 43, 12 772 (1991).

32. S. John and T. Quang, *Phys. Rev.* A 50, 1764 (1994).

33. S. Bay, P. Lambropoulos, and K. Molmer, *Opt. Commun.* 132 257 (1996).

34. S. John and T. Quang, *Phys. Rev. Lett.* 74, 3419 (1995).

35. T. Quang and S. John, *Phys. Rev.* A 56, 4273 (1997).

36. S. John and V.I. Rupasov, *Phys. Rev. Lett.* 79, 821 (1997)

37. S.L. McCall et al., *Phys. Rev. Lett.* 67, 2017 (1991).

38. W.M. Robertson et al., *Phys. Rev. Lett.* 68, 2023 (1992).

39. U. Gruning, V. Lehmann, and C.M. Engelhardt, *Appl. Phys. Lett.* 66, 3254 (1995).

40. U. Gruning, V. Lehman, S. Ottow, and K. Busch, *Appl. Phys. Lett.* 68, 747 (1996).

CHAPTER 3

PHOTONIC CRYSTALS

Chapter contents

3. 1 Introduction

The electronics revolution of the past 50 years has its roots in two scientific and technological areas. On the one hand, there have been tremendous advancements in our understanding of the physics of metals, dielectrics, and semi-conductors, leading to the development of devices such as the transistor. On the other hand, a variety of processing techniques such as thin-film growth and deposition, ion implantation, and photolithography have allowed the massive integration of electronic functionality within a very small area, leading to microprocessors and high-density memory, among other innovations. Our ability to control photons is in many ways in its infancy, compared with how we can manipulate electrons. Passive devices such as optical fibers, waveguides, splitters, and multiplexers are well developed. But as soon as more complex functionality or integration is required, the optical solutions do not yet exist. For example, all-optical switches are still very rudimentary and bulky, and the size of an optical integrated circuit (IC) is most often in the millimeter or centimeter range rather than the submicrometer dimensions common in electronic technology.

Nevertheless, there is a clear need to develop new materials and concepts with increased optical functionality for a variety of applications. The global telecommunications market is on an extraordinarily steep growth curve, driven largely by the explosion of the Internet, which plays an increasingly pervasive role in our daily life. The demand for broadband communications networks is expected to grow for many years to come. New approaches for the manipulation of photons will have to be developed to realize the more advanced optical elements needed for networks in the coming decade. Photonic crystals may play an important role in this development.

3.2 Naturally Occurring Photonic Crystals

A prominent example of a photonic crystal is the naturally occurring gemstone opal (Figure 3.1). Its play of colours is essentially a photonic crystal phenomenon based on Bragg diffraction of light on the crystal's lattice planes. Another well-known photonic crystal is found on the wings of some butterflies such as those of genus Morpho [1]. Photonic crystals in nature is detailed in chapter 7.

Figure 3.1: Naturally occurring photonic crystal

3.3 History of Acoustic and Photonic Crystals

Photonic crystals, also known as photonic band gap (PBG) structures, are structures with a periodically modulated dielectric constant. Main reason for the study of photonic crystals is that, in analogy to the case of an electron moving in a periodic potential, certain photon frequencies can become forbidden, independent of photon polarization and the direction of propagation - a complete photonic bandgap (CPBG). There is a common belief that, in the near future, photonic crystals systems will allow us to perform many functions with light that ordinary crystals do with electrons. At the same time, photonic structures are of great promise to become a laboratory for testing fundamental processes involving interactions of radiation with matter in novel conditions. A bandgap can also occur in acoustic crystal structures, which are structures charaterized by a periodically modulated elastic constant. Its obvious use is that of an acoustic isolator.

A study of wave propagation in periodic structures has a long history, which stretches back to, at least, Lord Rayleigh classical article on the influence of obstacles arranged in rectangular order upon the properties of a medium [1]. This work was soon followed by work of Kasterin [2], who studied reflection and refraction of low frequency sound by an orthorombic grid of hard spheres (an acoustic crystal). Another early paper on photonic crystals is that of Ewald [3]. Ewald studied if the optical birefringence can be caused by lattice structure alone, i.e. if it requires anisotropy of the atomic scatterers. Later on, wave propagation in periodic structures was a subject of the book [4] of the same title by Brillouin and Parodi. The use of two-dimensional distributed feedback periodic structures in thin film optical devices and lasers is disclosed in the US patent publication US-3,884,549 [5]. In the second half of seventies and early eighties a popular subject was an investigation of absorption spectrum of clusters of spheres. Within this subject, results were also obtained for regular finite and infinite linear chains of metal spheres [6]. Without knowing that, the authors basically rediscovered and applied Beeby's multiple-scattering theory [7] to electromagnetic waves. Some of early history of acoustic and photonic crystals can also be found in a review [8] by Korringa.

3.4 Photonic Crystal

Photonic crystals are periodic compounds of different dielectric media extending in one, two or three dimensions, capable of exhibiting a complete photonic bandgap, prohibiting the propagation of light in over a finite spectral range. One-dimensional (1D) photonic crystals e.g. are periodic assemblies of dielectric slabs with alternating dielectric constants. Light propagation in the direction normal to the stack is forbidden over a finite spectral range for any dielectric contrast between the two materials in the stack. If the periodicity is applied to two or three dimensions, a sufficiently large dielectric contrast is required before a complete photonic bandgap can be achieved in all directions of periodicity. For this reason, 2D and 3D photonic crystals are typically fabricated using semi-conductors, which possess high dielectric constants. Besides the dielectric contrast also the choice of the geometry of the periodicity is important. In contrast to a 1D photonic crystal, where the bandgap occurs due to Bragg scattering among the various slabs, in higher dimensions, a complete bandgap only occurs if the stop bands due to Bragg scattering overlap in frequency in all directions. It has been shown only for a few geometries to achieve this goal. The fabrication of such photonic crystals in the optical spectrum requires sophisticated micro fabrication techniques to produce high quality crystals with features well below one μm in size.

A photonic crystal is a regularly structured material that exhibits strong interaction with light. The conceptually simplest example of such a material is a multiplayer stack of alternating high- and low-dielectric-constant materials. Strong interaction with light occurs in such a material because of interference between the light beams that are reflected and refracted at all interfaces inside the material. The final optical response is determined by the coherent superposition of all of these optical waves. It has long been known that such multilayer stacks can be engineered to have, for example, nearly perfect reflection over a (narrow or broad) wavelength range, a so-called stop band [9]. Thin-film deposition techniques have made such structures widely available. Well-known examples of such "one-dimensional" (1D) photonic crystals are dielectric mirrors, filters, and fiber gratings, distributed-feedback structures and vertical-cavity surface-emitting lasers. Research is also being focused on "omni-directional" mirrors that reflect light over a well-defined wavelength range in all directions, again using an alternating array of thin films with appropriately chosen optical properties [10].

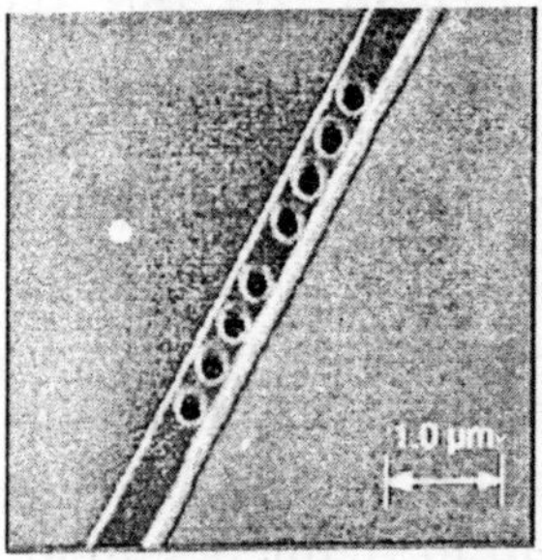

Figure 3.2: Scanning electron micrograph of a one-dimensional Silicon photonic crystal integrated in an optical channel waveguide.

Figure 3.2 shows an example of a 1D photonic crystal integrated in an optical channel waveguide [11]. In this structure, fully based on silicon, an array of holes was etched by using standard lithographic techniques. The size and spacing between the holes defines the wavelength-dependence of optical transmission through the waveguide. The importance of this structure lies in its extremely small size and mode volume, and the possibility of its integration on a planar (silicon) substrate.

While many of the 1D structures mentioned have a wealth of applications, the fundamental optical concept behind their operation is relatively simple. In recent years, there has been much activity aimed at expanding the simple concepts of layered, 1D photonic structures to higher dimensions. As first proposed by Yablonovitch [12] and John [13], the optical properties of such materials can be described by an "optical band structure." This concept has analogies to the well-known band structure of electronic materials in the sense that in material with particular structures, it predicts the existence of an optical bandgap, that is, a range of optical frequencies that cannot propagate in the material. This concept is particularly intriguing in a 3D photonic crystal, as it implies that in a particular frequency band, spontaneous emission would be completely suppressed. Initial searches for a structure that would possess a full bandgap led to fcc crystalline structures, but they failed to yield positive results. Soukoulis et al. made the important discovery that diamond symmetry eliminated degeneracy in the band diagram, thus opening up a bandgap in all crystal directions [14]. In the past few years, great progress has been made to realize such structures experimentally.

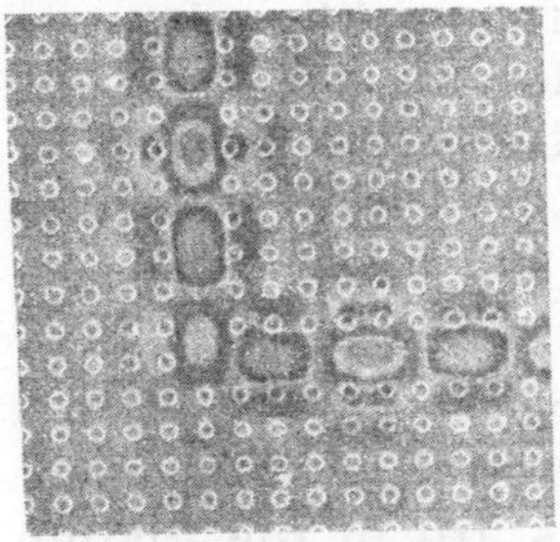

Figure3.3: Simulation of the propagation of an optical mode traveling around a sharply bent waveguide in a (two-dimensional) 2D photonic crystal composed of a cubic array of dielectric cylinders. The calculation is performed for a refractive index n= 3.4, and the spacing between the centers of the cylinders is 0.35the wavelength (e.g., at a wavelength of 1.55 μm, the spacing is 550 nm).

While 2D crystals seem less appealing because of their lack of optical control over the third dimension, they have the advantage of possible integration with planar optical-waveguide technology. In addition, external probes can be used to determine properties inside the crystal. The group led by Joannopoulos has proposed a large variety of 2D structures, many of which are now being studied experimentally.

As an example, Figure 3.3 shows a simulation of the propagation of an optical mode traveling around a sharply bent waveguide in 2D photonic crystal composed of a cubic array of dielectric cylinders. Note that this particular property of the photonic crystals is due to a carefully chosen defect, that is, a missing row and column of cylinders. Indeed, defects and disorder play an extremely important role in photonic-crystal research, as they enable the tailoring of particular properties for specific wavelengths. The first experimentalrealizations of 3D photonic crystals were for wavelengths in the microwave region. Yablonovitch and collaborators [15] invented an ingenious scheme of holes made in a dielectric using mechanical drilling (see Figure 3.4).

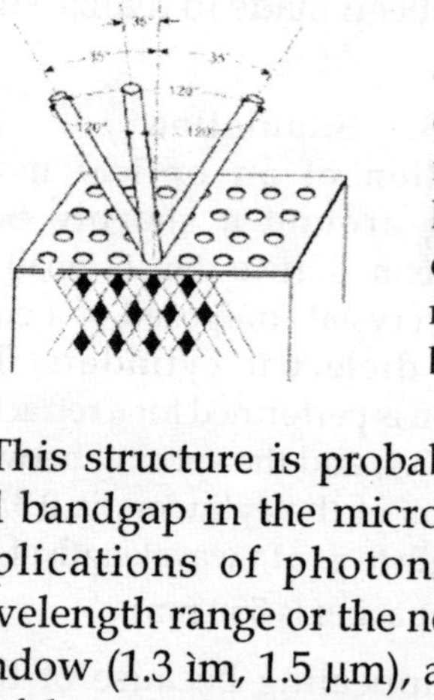

Figure 3.4: Schematic diagram of the construction of an fcc lattice. Three sets of holes are drilled on a 35 tilted axis into a slab of material. For microwave length scales, these holes can be drilled mechanically, whereas in the near-infrared, reactive ion etching has been used.

This structure is probably the first 3D structure with a full bandgap in the microwave regime. Obviously, many applications of photonic crystals are in the visible wavelength range or the near-infrared telecommunications window (1.3 ìm, 1.5 μm), and submicrometer resolution in the fabrication technology is therefore required. This can be done by taking advantage of the latest techniques in submicrometer patterning, initially developed for the IC industry.

As an example, Figure 3.5 shows 2D photonic crystal composed of Si pillars with diameters as small as 205 nm made using high-resolution lithography [16]. Alternative methods such as the self-assembly of colloids have led to important new photonic-crystal fabrication technology as well.

Figure 3.5: Scanning electron micrograph of a periodic array of silicon pillars fabricated using deep anisotropic etching. The silicon pillars are 205 nm in diameter and 5μm tall. This structure possesses a bandgap of around 1.5 μm for transverse magnetic polarization. By removing an array of pillars, a waveguide bend may be fabricated. Input and output waveguides are integrated with the photonic crystal.

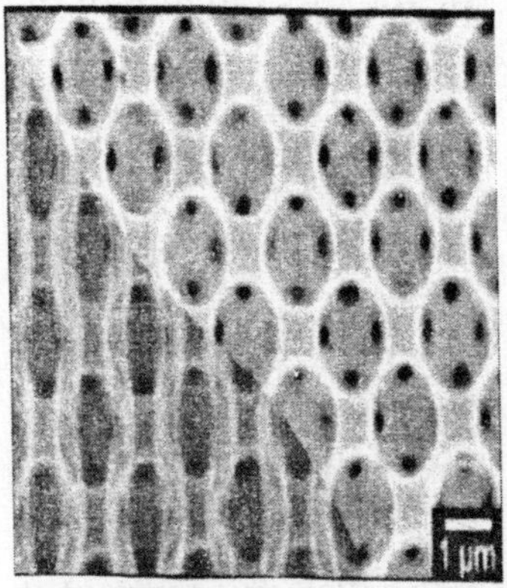

Figure 3.6. A scanning electron micrograph (SEM) of 3D photonic crystal. This structure was assembled by sedimentation of monodisperse colloidal silica on a template, thus forming an fcstructure Subsequently, molten selenium, which has an index of refraction of 2.5 in the near-infrared and very low absorption, was imbibed into the interstitial space, and the silica was etched away to produce a high-dielectric constant replica.

A promising example is shown in Figure 3.6. A completely different approach to building 3D structures with micrometer scale features was demonstrated by Marder and collaborators.

Semi conductors such as Si and GaAs possess the high dielectric contrast and low absorption required for a full photonic bandgap in two or three dimensions. At the same time, photonic crystals made of dielectric materials with a lower refractive index, such as SiO_2, TiO_2 and polymers, while not having a full bandgap, can still have strong interaction with light and, therefore, interesting photonic properties. More recently, theory on photonic crystals partly composed of metals indicates a wealth of interesting phenomena in such materials, and the first experiments in this area are just appearing [17-19].

3.5 Cavity Resonators

Photonic crystals (or photonic bandgap structures = PBG) are machined materials with a periodic spatial variation of the dielectric constant. Due to Bragg reflection electromagnetic waves of certain frequency ranges cannot propagate through the photonic crystal and resonant modes cannot exist. These frequency intervals are called photonic bandgaps. In a dielectric loaded cavity resonator, the quality factor of a resonant mode is mainly determined by two loss mechanisms:

1. Dielectric losses determined by the loss tangent of the employed dielectric material
2. Metal losses due to surface currents in the metallic cavity walls

Figure 3.7: A simulation of a whispering-gallery mode resonance in an extended defect in a 2 dimensional photonic crystal consisting of air holes in dielectric material with a dielectric constant of 10.

Whispering-gallery type (Figure 3.7) extended defect modes in photonic crystals have been investigated, where the electromagnetic energy of the resonant mode is stored to a large extend in air, i.e. the dielectric filling factor (ratio of electric field energy stored in the dielectric material to total electric field energy) is very small. Similar to dielectric resonators, metal losses in the walls of the embedding shielding cavity is very small. Such resonators should allow for very high Q-values from the microwave to the infrared frequency range. As an example, high-Q resonators at micro- and millimeter waves can be used for low-phase-noise oscillators in advanced microwave communication and radar systems. A resonator setup for a 3D TE mode microwave cavity with a 2D photonic crystal lattice and a schematic side view of resonator geometry is shown in figures 3.8and 3.9.

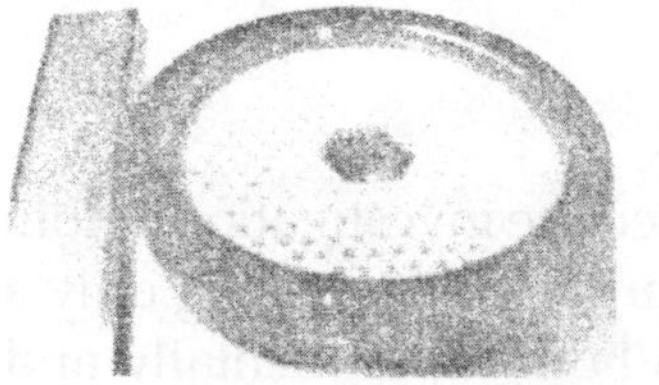

Figure.3.7: Resonator setup for a 3D TE mode microwave cavity with a 2D photonic crystal lattice.

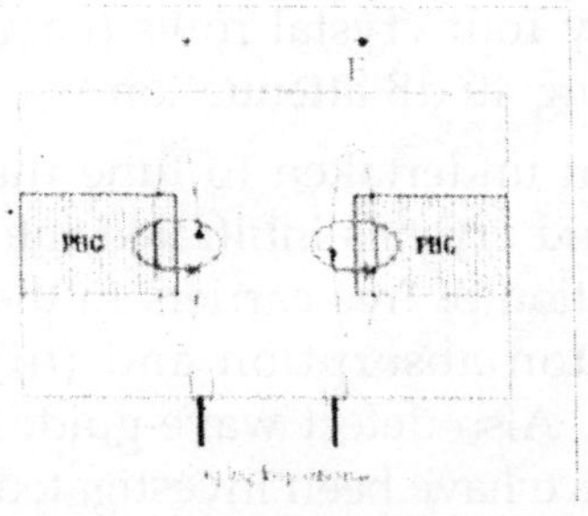

Figure.3.8: Schematic side view of the resonator geometry and field lines in the cavity.

3.6 Two-Dimensional Photonic Crystals

3.6.1 Two-Dimensional Photonic Crystals in Silicon

The optical properties of 2D photonic crystals made of a triangular lattice of air holes in silicon (known as macro porous silicon) are experimentally studied. The samples are fabricated at the Max Planck Institute for Microstructure Physics in Halle, Germany, by the research groups of U. Göesele and R.B. Wehrspohn. The crystals possess a complete photonic bandgap (i.e. a stop band in all directions within the 2D plane for both polarizations of light) at a wavelength of approximately 4 microns. The optical characterization of bulk crystals is performed using Fourier transform infrared (FTIR) spectroscopy in both Halleand Toronto.

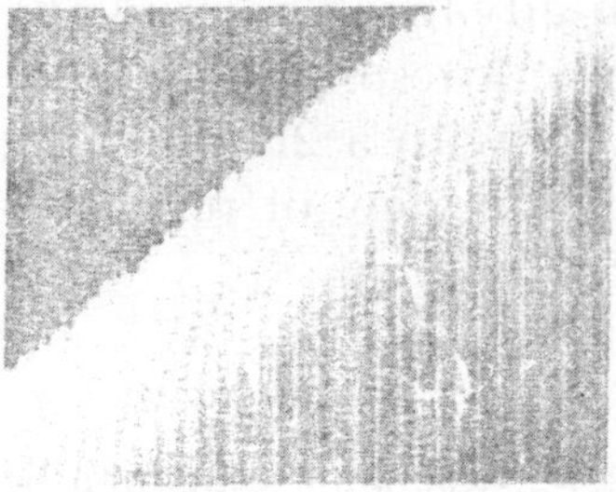

Figure. 3.10: 2-D photonic crystals made of a triangular lattice of air holes in silicon.

Successful projects have involved the investigation of thin 2D silicon photonic crystals (Figure 3.10) possessing only a few crystal layers. It was shown both experimentally and theoretically that the attenuation per crystal row is approximately 10 dB, with only four crystal rows (i.e. a crystal 6 microns thick) exhibiting 40 dB attenuation.

More recently work has been undertaken to tune the photonic bandgap using (i) liquid crystals infiltrated into the photonic lattices, (ii) generation of free carriers in the silicon via one- and two-photon absorption and (iii) non-resonant Kerr non-linearity. Also defect wave-guides patterned with the photonic lattice have been investigated for their transmission properties. The experiment demonstrated the principle of guiding light by the Bragg scattering properties of the photonic crystal.

3.6.2 Two-Dimensional Photonic Crystals in Gallium Arsenide

GaAs material system possesses non-linearity larger than that of Silicon and is ideal for ultra fast non-linear experimentation with photonic crystal. The fabrication techniques employed by the semi-conductor community are well developed and are readily transferable to optoelectronics and photonic crystals.

A square lattice of air holes patterned into GaAs siabs is fabricated by electronic beam lithography and reactive ion etching. Non-linear experiments have taken place observing strong enhancement of the second harmonic generation in the photonic crystal. This effect is a consequenence of the photonic band structure and is 1000 times larger than that observed in unpatterned GaAs slabs.

3.7 Three-Dimensional Photonic Crystals

A complete 3D photonic bandgap in a self-organized photonic crystal is shown in Figure 3.11. A face centered cubic lattice of air spheres in silicon, commonly known as an inverse opal, was fabricated at the University of Toronto by a team of researchers including A. Blanco, E. Chomski, S. Grabtchak, M. Ibisate, S.John, S.W. Leonard, C. Lopez, F. Meseguer, H. Miguez, J.P. Mondia, G.A. Ozin, O. Toader, and H.M. van Driel. Chemical vapour deposition was used to deposit silicon inside an artificial opal, which is a face centered cubic lattice of silica spheres.

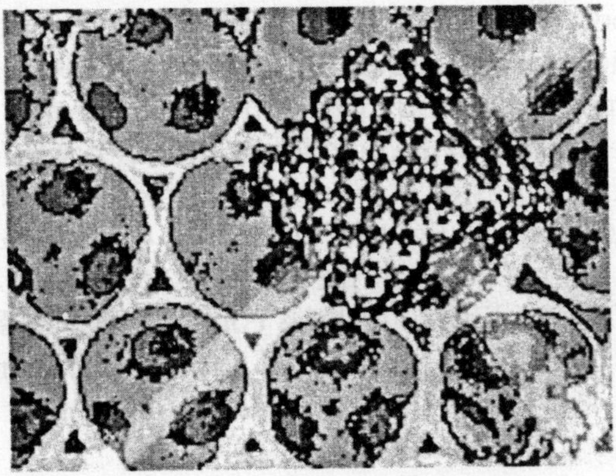

Fig. 3.11: 3D photonic bandgap in a self-organized photonic crystal

The artificial opals were fabricated in Spain at the Institute de Ciencia de Materiales de Madrid (CSIC) and Unidad Asociada (CSIC-UPV) Universidad Politécnica. The infiltrated structure consisted of a connected network of silica spheres coated with silicon. Although this structure has many interesting optical properties, it does not possess sufficient refractive index contrast to exhibit a complete photonic bandgap. The silica spheres were therefore chemically removed by etching the crystal in HF, leaving behind the inverse opal structure. Based on optical reflectivity measurements and calculated band structures, inferred the presence of a complete photonic bandgap at a wavelength of 1.5 microns.

3. 8 Photonic Crystal Fibers

Photonic-crystal fiber (PCF), also spelled *fibre*, is a new class of optical fiber based on the properties of photonic crystals. Because of its ability to confine light in hollow cores or with confinement characteristics not possible in conventional optical fiber, PCF is now finding applications in fiber-optic communications, fiber lasers, non-linear devices, high-power transmission, highly sensitive gas sensors, and other areas. The term "photonic-crystal fiber" was coined by Philip Russell in 1995-1997 (he states (2003) that the idea dates to unpublished work in 1991), although other terms such as microstructured fiber are also used and the nomenclature in the field is not entirely consistent. More specific categories of PCF include photonic-bandgap fiber (PCFs that confine light by band gap effects), holey fiber (PCFs using air holes in their cross-sections), *hole-assisted* fiber (PCFs guiding light by a conventional higher-index core modified by the presence of air holes), and Bragg fiber (photonic-bandgap fiber formed by concentric rings of multilayer film)[20-22].

In general, such fibers have a cross-section (normally uniform along the fiber length) microstructured from two or more materials, most commonly arranged periodically over much of the cross-section, usually as a "cladding" surrounding a core (or several cores) where light is confined. For example, the fibers first demonstrated by

Russell consisted of a hexagonal lattice of air holes in a silica fiber, with a solid (1996) or hollow (1998) core at the center where light is guided. Other arrangements include concentric rings of two or more materials, first proposed as "Bragg fibers" by Yeh and Yariv (1978), a variant of which was recently fabricated by Temelkuran *et al.* (2002).

(Note: PCFs and, in particular, Bragg fibers, should not be confused with fiber Bragg gratings, which consist of a periodic refractive index or structural variation along the fiber axis, as opposed to variations in the transverse directions as in PCF. Both PCFs and fiber Bragg gratings employ Bragg diffraction phenomena, albeit in different directions.)

Generally, such fibers are constructed by the same methods as other optical fibers: first, one constructs a "preform" on the scale of centimeters in size, and then heats the preform and draws it down to a much smaller diameter (often nearly as small as a human hair), shrinking the preform cross-section but (usually) maintaining the same features. In this way, kilometers of fiber can be produced from a single preform. Most photonic crystal fiber has been fabricated in silica glass, but other glasses have also been used to obtain particular optical properties (such has high optical non-linearity). There is also a growing interest in making them from polymer, where a wide variety of structures have been explored, including graded index structures, ring structured fibres and hollow core fibers. These polymer fibers have been termed "MPOF", short for microstructured polymer optical fibers (van Eijkelenborg, 2001). A combination of a polymer and a chalcogenide glass was used by Temelkuran *et al.* (2002) for 10.6 μm wavelengths (where silica is not transparent).

Photonic crystal fibers can be divided into two modes of operation, according to their mechanism for confinement.

Those with a solid core, or a core with a higher average index than the microstructured cladding, can operate on the same index-guiding principle as conventional optical fiber – however, they can have a much higher effective-index contrast between core and cladding, and therefore can have much stronger confinement for applications in non-linear optical devices, polarization-maintaining fibers (or they can also be made with much *lower* effective index contrast). Alternatively, one can create a "photonic bandgap" fiber, in which the light is confined by a photonic bandgap created by the microstructured cladding – such a bandgap, properly designed, can confine light in a *lower-index* core and even a hollow (air) core. Bandgap fibers with hollow cores can potentially circumvent limits imposed by available materials, for example to create fibers that guide light in wavelengths for which transparent materials are not available (because the light is primarily in the air, not in the solid materials). Another potential advantage of a hollow core is that one can dynamically introduce materials into the core, such as a gas that is to be analyzed for the presence of some substance.

The periodic arrangement of ions on a lattice gives rise to the energy band structure in semi-conductors. Energy bands control the motion of charge carriers through the crystal. Similarly, in a photonic crystal, the periodic arrangement of refractive index variation, controls how photons are able to move through the crystal. Replacing the ions on a lattice are regions of low refractive index within a high-refractive index material or vice versa. Photons react to the refractive index contrast in an analogous manner to the way electrons react when confronted with a periodic potential of ions. Each results in a range of allowed energies and a band structure characterized by an energy gap or photonic bandgap. A bandgap forms when the electron wavelength is comparable to the inter-atomic spacing.

Eli Yablonovitch and Sajeev John first predicted photonic bandgaps. An array of 1 mm holes milled into a slab of material of refractive index 3.6, latter to be known as Yablonovitch, was found to prevent microwaves from propagating in any direction. An estimate of the distance between voids is given by the wavelength of the light divided by the refractive index of the material. This relationship means that it is even more difficult to create photonic crystals in materials with high refractive index. Finding materials and techniques of producing photonic crystals has remained illusive until 1997.

Breaking the periodicity of the voids in the photonic crystal, either by enlarging or reducing the size of a few of the voids, introduces new energy levels within the photonic bandgap. This is analogous to the creation of energy levels within the bandgap by the addition of dopant atoms in semi-conductor crystals. Narrow line width lasers [19], that are so important in the area of Dense Wavelength Division Multiplexing (DWDM) communication systems, can be fabricated using photonic crystals formed from the III-V semi-conductors or rare-earth doped glass. The defect mode or microcavities formed by breaking the periodicity of the crystal amplify only those wavelengths of light those are able to pass freely through the crystal. The wavelength range of emission is related to the diameter of the microcavity divided by the diameter of the regular holes. Photonic crystal microcavities are more efficient than conventional semi-conductor diode lasers since there are few directions in which the photons can escape. Passive devices, such as optical filters, where it is often required to filter a weak signal, the aim is to minimize further loss. Unfortunately, these losses are greater in materials with a high refractive index and photonic crystals use the contrast in refractive index to create energy bandgaps. Altering the geometry of the voids, in such a way as to increase the symmetry of the periodic structure, the bandgap can be restored in a low refractive index material. For example, by going from a rectangular array of voids in which the spacing of the holes is different for the two perpendicular directions within the plane, to a triangular arrangement which has a sixfold degree of symmetry within the plane. A quasiperiodic arrangement of holes has a photonic bandgap whatever the angle of the light travels.

Photonic crystals are also being exploited in the technology of optical fibers by Phillip Russell and Jonathan Knight at the University of Bath to create fibers that are carry light at high powers. Traditional optical fibers have a glass core surrounded by a cladding layer. The cladding has a lower refractive index than the core. The difference in the refractive index between the two materials is slight, but it is enough to cause light within the core to be reflected along the core by total internal reflection. At high powers, the light is concentrated in a very small region of space. Raman scattering can result, causing the light signal to be corrupted and at high powers the fiber can even be damaged.

The first photonic crystal fibers used a hexagonal array of air holes running the entire length of the fiber (Figure 3.11 and 3.12). The light traveled close to the fiber axis and so the power was still limited. The central tubes are removed to leave central holes of 15 mm diameter. The hole acts as a waveguide, the light being trapped by the photonic crystal structure around the hole. The photonic crystal allows only some of the light to propagate along the fiber.

Figure 3.12: SEM micrographs of a photonic-crystal fiber produced at US Naval Research Laboratory. (left) The diameter of the solid core at the center of the fiber is 5 μm, while (right) the diameter of the holes is 4 μm

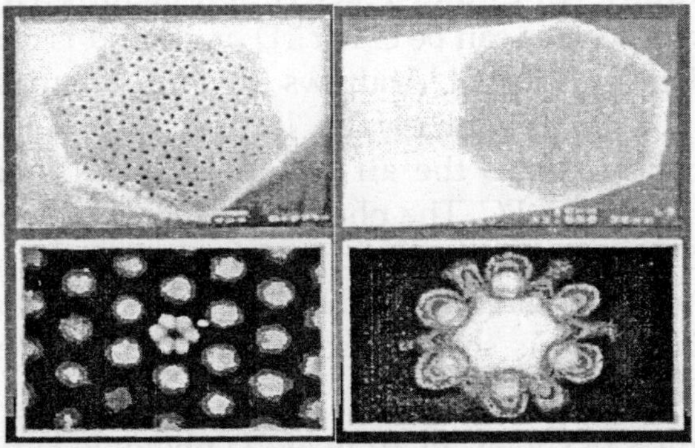

Figure 3.13: Holey fibre

Photonic crystal fiber is made by stacking pure silica tubes in a hexagonal array, and then heating and stretching it. This reduces the width of the structure, while retaining the geometry.

Ultimately, Professor Russell sees hollow-core fibers as the basis of ultra high-power fiber-delivery systems, in which light from a powerful laser source is delivered to a cutting head in surgery and machining. And one day soon, hollow-core fiber might guide atoms or particles over long distances, forming the basis of fiber-optic devices that move matter.

3.9 Photonic Crystals Waveguide

Photonic crystal waveguides (PCWs) in GaAs material systems can provide functionality into optoelectronic circuits and devices. A device can be passive or activated with quantum wells or quantum dots. The planar waveguides are grown at the EPSRC III-V Central Facility (University of Sheffield) by means of molecular beam epitaxy (MBE) or metal-organic vapour phase epitaxy (MOVPE). The waveguide provides vertical confinement of the electromagnetic (EM) field to a region that can be etched to form the photonic crystal (PC). A periodic lattice is etched from the front surface with electron beam lithography (EBL) and reactive ion etching (RIE). The etching processes can be either 1D or 2D and is carried out at the University of St. Andrews (by Prof. Thomas Krauss' group). The period of the lattice and the ratio of semi-conductor to the air channels/holes lead to the formation of the PC. The photonic dispersions are defined by a photonic Brillouin zone, analogous to that for electronic Brillouin zone at the edge of the electronic band structure. This is selected so as to have an operational wavelength of 700-1100 nm.

Figure 3.13 shows an array of 2D photonic lattices with varying periods under an optical microscope. The different periods produce unique photons dispersion relationships resulting in different wavelengths of the stop band.

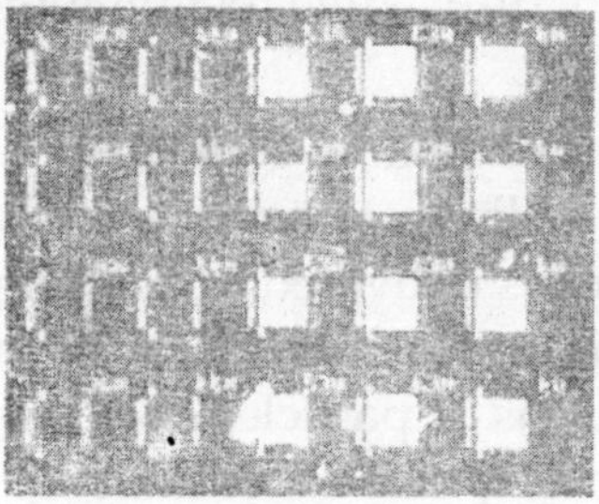

Figure 3.14: An array of photonic bandgap crystals with varying periods. The lattice period increases from left to right. This picture was taken under an optical microscope with a white light source incident at approximately 45° to the normal.

The introduction of a photonic lattice can confine the light in the in-plane direction. Figure 3.15(c) shows a single unit cell for a 1D etched waveguide. The light confinement is shown for the two photonic modes that are supported for this specific structure under reflectivity measurement at 15°. One-dimensional passive samples are relatively easy to interpret with respect to the photonic band structure as only one direction has a photonic Brillouin zone to investigate [22-24]. This can be achieved with a surface coupling, angle dependent, reflectivity measurement that has been perfected by the group. Using a spectrometer to scan the energy range we can build a picture for various values of the in-plane wavevector by varying the incident beam angle.

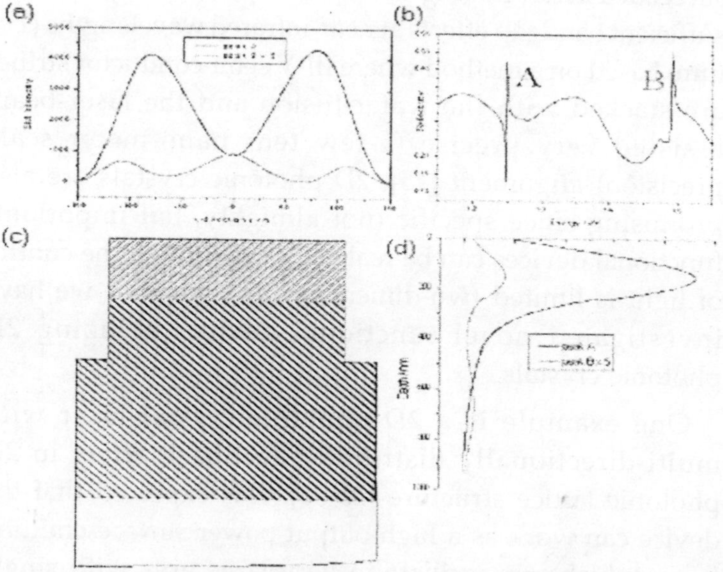

Figure 3.15: Calculations of the electromagnetic density for a single unit cell of a 1D - PC. The inplane density is shown in (a) the depth density is shown in (d) they are in scale with the schematic of the unit cell (c) and each plot shows two modes which have their energies defined from the reflectivity spectra in (b).

3.10 Photonic Crystals: Novel Nano-structures for Light

Much interest has been drawn in photonic crystals in which the refractive index changes periodically. A photonic bandgap is formed in the crystals, and the propagation of electromagnetic waves is prohibited for all wave vectors.Various important scientific and engineering applications such as control of spontaneous emission, zero-threshold lasing, very sharp bending of light, trapping of photons, and so on, are expected by utilizing the photonic bandgap and the artificially introduced defect states and/or light-emitters. Recently, scientists succeeded in developing complete 3D photonic crystals with sufficient bandgap effects at near-infrared wavelengths (1~2 mm) based on a method where III-V semi-conductor stripes are stacked with the wafer-fusion and the laser-beam assisted very precise (a few tens nano-meter scale precision) alignment [25]. 2D photonic crystals are also promising since specific (not almighty, but important) functional devices can be realized even though the control of light is limited two-dimensionally. Thus far, we have investigated novel functional devices utilizing 2D photonic crystals.

One example is a 2D photonic crystal laser with multi-directionally distributed feedback effect in 2D photonic lattice structure [26-30]. It is expected that the device can work as a high-output power surface-emitting laser, which can oscillate in very large area with single mode and desired polarization. The other is a device utilizing a single defect in 2D photonic bandgap structure. The defect traps the photons, which propagate through a waveguide, formed in 2D photonic crystal slab and emit them to free-space. This phenomenon is very promising for the actual application to an ultra-small optical device with a function of dropping (or adding) photons with various energies from (or into) optical communication traffic (fiber). There are many other important applications by utilizing the strong localization of photons at the defect such as enhancement of non-linear optical phenomena, and trapping of nanoparticles.

3.11 Tunable Photonic Bandgap Crystal

Researchers around the world are investigating the properties of photonic crystals, the low-loss periodic dielectric lattices that influence the propagation of electromagnetic waves. Semi-conductor materials with electronic bandgaps - electron energies that is forbidden within the crystal - were the cornerstones of the electronics revolution. Now, materials with electromagnetic bandgaps - photon energies that cannot propagate through the crystal - promise to lead a photonics revolution. Recently Dr. Sajeev John and his team at the University of Toronto developed a photonic crystal with a tunable bandgap. They infiltrated silicon into the air gaps of the crystal, and then etched out the opal, leaving silicon in an inverse opal crystal structure. The resultant periodic dielectric array exhibits an 8 percent bandgap at 1.5 μm; i.e., wavelengths between about 1.38 and 1.62 μm cannot propagate through the crystal. To achieve tunability, John coated the internal surfaces of the crystal with BEHA, a low-index nematic liquid crystal [Figure 3.16]. The liquid crystal reduces the bandgap to 1.6 per cent, but it can now be completely closed via an external magnetic field. This tunability is the key.

John said that the application of an electric field shifts the bandgap and some features of the band structure. He compared this to what happend in a semi-conductor when an electric field is applied: Shifting the electronic bands controls the flow of electrical current. In the tunable photonic bandgap, light can be steered with control of the material's linear electro-optic effect. The tunable bandgap could be used not only for simple on/off switching, but also for more localized control of the propagation. By applying electric fields to different regions within the crystal, the path of the light can be precisely controlled. John's discovery is sparking interest in the photonic crystal community.

Eli Yablonovitch commented on it - not only because of the tunability, but also because of the prospect that these materials may be easily manufactured. He said that the inverse silicon opal and the infiltration of the liquid crystal are straightforward. According to John, only two challenges remain: precisely controlling the degree of infiltration and ensuring that the liquid crystal wets the inner surfaces uniformly. If the low-cost and high-performance expectations are met, this could be another step in the photonics revolution.

3.12 Liquid-Crystal Photonic-Bandgap Materials

When an optically birefringent nematic liquid crystal is infiltrated into the void regions of an inverse opal, photonic-band-gap (PBG) material, the resulting composite material exhibits a completely tunable PBG. In particular, the three-dimensional PBG can be completely opened or closed by applying an electric field which rotates the axis of the nematic molecules relative to the inverse opal backbone. Tunable light localization effects may be realized by controlling the orientational disorder in the nematic (shown in Figure 3.16).

Figure. 3.16: Cross-sectional view through the inverse opal backbone (blue) resulting from incomplete infiltration of silicon into the air voids of an artificial opal. After etching out the template, a fcc lattice of overlapping air spheres remains and additional air voids appear as triangular or diamond shaped holes on the surface of the cut. A tunable PBG is obtained by infiltrating this backbone with nematic liquid crystal (yellow) which wets the inner surface of each sphere (only one is shown in the figure).

(PBG) materials facilitate two fundamentally new optical principles, namely the localization of light [24 -28] and the controllable inhibition of spontaneous emission of light from atoms and molecules [29-33]. The utility of PBG materials arises essentially from their ability to realize these two functions. Although an intensive effort has developed over the past ten years to microfabricate PBG materials, it is only recently that a clear route to synthesizing large scale three-dimensional PBG materials with submicron lattice constants has been demonstrated using self-assembly methods. The application of this approach to Si, Ge, and GaAs based PBG materials may open the door to applications in laser devices and telecommunications as well as to the realization of fundamentally new effects in quantum and non-linear optics.

For many applications, however, it is advantageous to obtain some degree of tunability of the photonic band structure through electro-optic effects. Such tunability may be obtained by controlling one or several forms of optical anisotropy of the constituent materials. The science of liquid crystals has spawned an entire industry related to these electro-optic effects [30-40]. In earlier works, however, a rather pessimistic conclusion regarding the efficacy of birefringent photonic crystals was drawn [40, 41] since attention was restricted to unrealizable structures that consist of spheres of disconnected anisotropic, high dielectric materials in an air background. In other morereadily realizable liquid crystal systems [39, 40], there was insufficient refractive index contrast to open a three-dimensional PBG. However, it is now well known that inverse opal structures, i.e., air inclusions in a high dielectric backbone material such as silicon provide a much more efficient scattering system both for ordered as well as disordered structures [41]. For these inverse opals the optimal filling ratios of the high dielectric backbone lies around 24.5 per cent, leaving a large empty volume for infiltration by a low refractive index liquid crystal with strong optical anisotropy. This large volume of birefringent material makes the resulting composite system highly efficacious for electro-optic tuning effects. In particular, a change in the orientation of the nematic director field with respect to the inverse opal backbone by an external electric field can completely open or close the full, three-dimensional PBG. In this Letter, we illustrate the principle of the fully tunable electromagnetic vacuum with detailed results on an inverted opal [42] made of silicon into which the low index nematic liquid crystal BEHA is infiltrated. For this structure we find that a 2 per cent photonic band gap may be opened or closed, by changing the orientation of the nematic director. In addition, the thermally driven nematic to isotropic liquid phase transition provides further tunability of the PBG. Localized states of light may be created through random fluctuations as well as through singularities (textures) in the nematic director field [43-47].

3.13 References

1. S. Kinoshita, S. Yoshioka and K. Kawagoe"Mechanisms of structural colour in the Morpho butterfly: cooperation of regularity and irregularity in an iridescent scale" *Proc. R Soc. Lond.* B 269, 1417-1421 (2002)
2. Lord Rayleigh, On the influence of obstacles arranged in rectangular order upon the properties of a medium, *Philos. Mag.* **34**, 481-502 (1892).
3. N. P. Kasterin, *Vers. Akad. van We :nschappen Amsterdam* **6**, 460 (1897-8); Diss. Moscow (1903).
4. P. P. Ewald, Zur Begrundung der Kristalloptik, Ann. *Phys.* (Lepzig) **49**, 1-38 (1916).
5. L. Brillouin and M. Parodi, *Wave Propagatic ı in Periodic Structures* (Dover, New York, 1953).
6. S. Wang and S. K. Sheem, *Two dimensional distributed feedback devices and lasers,* US patent publication US-3,884,549 A (May 20, 1975).
7. J. M. Gerardy and M. Ausloos, Absorption spectrum of clusters of spheres from the general solution of Maxwell's equations. II. Optical properties of aggregated metal spheres, *Phys. Rev.* B **25**, 4204-4229 (1982).
8. J. L. Beeby, The electronic structure of disordered systems, *Proc. Roy. Soc.* A **279**, 82-97 (1964).
9. J. Korringa, Early history of multiple scattering theory for ordered systems, *Phys. Rep.* **238**, 341-60 (1994).
10. D. Joannopoulos, R.D. Meade, and J.N. Winn, photonic Crystals (Princeton University Press,Princeton, NJ, 1995).
11. M.F. Weber, C.A. Stover, L.R. Gilbert, T.J.Nevitt, and A.J. Ouderkirk, *Science* 287 (2000) p. 2451.
12. S. Foresi, P.R. Villeneuve, J. Ferrera, E.R. Thoen, G. Steinmeyer, S. Fan, J.D. Joannopoulos, L.C. Kimerling, H.I. Smith, and E.P. Ippen, *Nature* 390 (1997)

13. E. Yablonovitch, *Phys. Rev. Lett.* 58 (1987) p. 2059; E. Yablonovitch and T.J. Gmitter, *Phys. Rev. Lett.* 63 (1989) p. 1950.

14. S. John, *Phys. Rev. Lett.* 58 (1987) p. 2486.

15. K.M. Ho, C.T. Chan, and C.M. Soukoulis, Phys. Rev. Lett. 65 (1990)

16. E. Yablonovitch, T.J. Gmitter, and K.M. Leung, *Phys. Rev. Lett.* 67 (1991)

17. T. Zijlstra, E.W.J.M. van der Drift, M.J.A. de Dood, E. Snoeks, and A. Polman, J. Vac. Sci. Technol., B 17 (1999) p. 2734.

18. B.H. Cumpston, S.P. Ananthavel, S. Barlow, D.L. Dyer, J.E. , L.L. Erskine, A.A. Heikal, S.M. Kuebler, I.Y.S. Lee, D. McCord-Maughon, J.Q. Qin, H. Rockel, M. Rumi, X.L.Wu, S.R. Marder, and J.W. Perry, *Nature* 398(1999)

19. Hirayama, Hamano and Aoyagi, *Appl. Phys. Lett* 69 (6) August 1996, p791-793.

20. P. St. J. Russell, "Photonic crystal fibers," *Science* **299**, 358-362 (2003). (Review article.)

21. P. St. J. Russell, "Photonic crystal fibers", *J. Lightwave. Technol.*, **24** (12), 4729-4749 (2006). (Review article.)

22. F. Zolla, G. Renversez, A. Nicolet, B. Kuhlmey, S. Guenneau, D. Felbacq, *Foundations of Photonic Crystal Fibres* (Imperial College Press, London, 2005).

23. Joanopoulos, Meade, Winn, *Photnic Crystals; Molding the low of Light* (Princeton Uni. Press, 1995)

24. V.N. Astratov et al., "Heavy photon dispersions in photonic waveguides" *Applied Physics Letters* 77(2) 178, (2000).

25. S. Noda, K. Tomoda, N.Yamamoto, and A. Chutinan, "Full Three-Dimensional photonic Bandgap Crystals at Near-infrared Wavelengths", *Science*, 289, 604-606 (2000).

26. Imada, S. Noda, et a., "Coherent Two-Dimensional LasingAction in Surface-Emitting Laser with Triangular -Lattice Photonic Crystal Structure", *Appl.Phy.,Lett.*, 75, 316-318(1999).

27. Noda, M.Yokoyama, et al.,"Polarization Mode Control of Two-Dimensional Photonic Crystal Laser by Unit Cell Structure Design", *Science*, 293, 1123-1125 (2001).

28. S. John, *Phys. Rev. Lett.* 53, 2169 (1984).

29. S. John, *Phys. Rev. Lett.* 58, 2486 (1987).

30. P. W. Anderson, *Philos. Mag.* B 52, 505 (1985).

31. A. Z. Genack and N. Garcia, *Phys. Rev. Lett.* 66, 2064 (1991).

32. D. Wiersma et al., *Nature* (London) 390, 671 (1997).

33. V. P. Bykov, *Sov. J. Quantum Electron.* 4, 861 (1975).

34. E. Yablonovitch, *Phys. Rev. Lett.* 58, 2059 (1987).

35. S. John and J. Wang, *Phys. Rev. Lett.* 64, 2418 (1990).

36. S. John and T. Quang, *Phys. Rev.* A 50, 1764 (1994).

37. T. Quang et al., *Phys. Rev. Lett.* 79, 5238 (1997).

38. P. G. de Gennes and J. Prost, T*he Physics of Liquid Crystals* (Clarendon Press, Oxford, England, 1993).

39. S. Chandrasekhar, *Liquid Crystals* (Cambridge University Press, Cambridge, England, 1992).

40. L. M. Blinov and V. G. Chigrinov, *Electro-Optic Effects in Liquid Crystal Materials* (Springer, New York, 1994).

41. Z.-Y. Li, J. Wang, and B.-Y. Gu, *Phys. Rev.* B 58, 3721(1998).

42. R. M. Hornreich, S. Shtrikman, and C. Sommers, *Phys.Rev.* E 47, 2067 (1993).

43. V. I. Kopp, B. Fan, H. K. M. Vithana, and A. Z. Genack, *Opt. Lett.* 23, 1707 (1998).

44. H. S. Sözuer, J. W. Haus, and R. Inguva, *Phys. Rev.* B 45, 13 962 (1992).

45. K. Busch and C. M. Soukoulis, *Phys. Rev. Lett.* 75, 3442(1995).

46. J. E. G. J. Wijnhoven and W. L. Vos, *Science* 281, 802(1998).

CHAPTER 4

PHOTON LOCALISATION AND APPLICATIONS

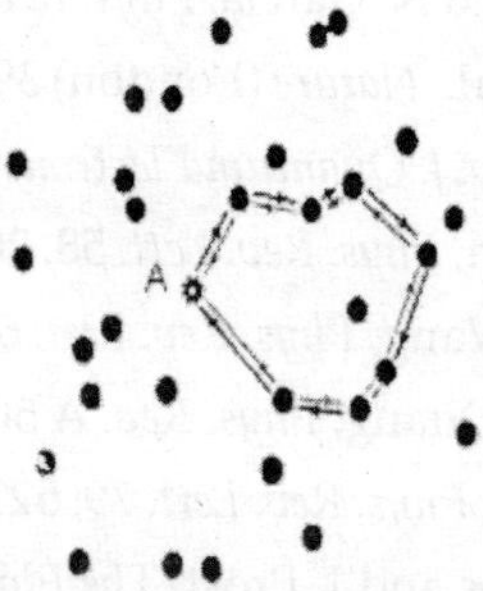

Chapter contents

4. 1 Introduction

Photonic lattices are materials specially designed to cage light. By shaking the mesh of the cage, we can see how light, initially passing freely, finds its room for maneuver progressively restricted. Photon localization, in three-dimensional dielectric media, opens a new frontier for fundamental phenomena in classical and quantum electrodynamics. Photon localization is a fundamental effect predicted to occur in certain strongly scattering dielectric microstructures. The realization of this effect has been facilitated by the prediction [1, 2] and development [3-5] of PBG materials. These are lossless materials, which exhibit a range of frequencies for which electromagnetic wave propagation is classically forbidden.

The idea of a photonic crystal – a structure explicitly designed to trap light in the same way that electrons are trapped in a semi-conductor. In photonic crystals, the refractive index varies in a periodic fashion, causing multiple light scattering and thus destructive interference. This effect prevents light from flowing in a range of frequencies known as the photonic bandgap. In this gap, the usual features of electromagnetism, as described by Maxwell's equations, can be partially or completely suspended. The resulting microscopic control of photons is even more remarkable than the control of electrons in microelectronic circuits based on semi-conductor technology.

In outer space, with no scattering medium, it travels for billions of years and allows us to glimpse deep into the Universe's antiquity. In the past few decades, it has become apparent that light can also be confined within very small spaces. Out in the cosmos, this occurs in black holes, in which light is restricted to a sphere of just a few kilometres' radius by an immense gravitational field. Back on Earth, fibre optics exploits the fact that light can be confined in two dimensions by internal reflection, while propagating freely in the third. Constructs known as laser cavities are used to trap light through mirror reflections, allowing very high light intensities to build-up.

In a perfect atomic crystal, visible light travels ballistically, only occasionally scattering off a lattice atom. As disorder increases—for example, through the introduction of a foreign atomic species—so does the frequency of scattering, and the light's passage through the crystal becomes diffusive. Ultimately, if the disorder is such that interference occurs between individual scattering events, then light can become spatially localized. Philip Anderson first introduced this kind of localization in 1958, for electrons and their spins confined in a semi-conductor.

Anderson went on to pioneer the field of light localization in the 1980s, at the same time as Sajeev John. These two researchers' theoretical predictions were followed by the observation of a weak form of light localization using coherent backscattering. A stronger form of light localization has been observed more recently in highly scattering, insulating (dielectric) media, in which the deviation from classical diffusion was measured by transmission spectroscopy.

Anderson localization brings classical diffusion to a complete halt. That is, on increasing the amount of scattering beyond a critical value, the material makes a transition into a localized state (Figure 4.1). This transition can best be observed in the transmission properties of the system. In the localized state, the transmission coefficient decreases exponentially instead of linearly with the thickness of a sample. At the transition, the transmission coefficient is expected to have a power-law dependence on the inverse thickness, which is probably quadratic.

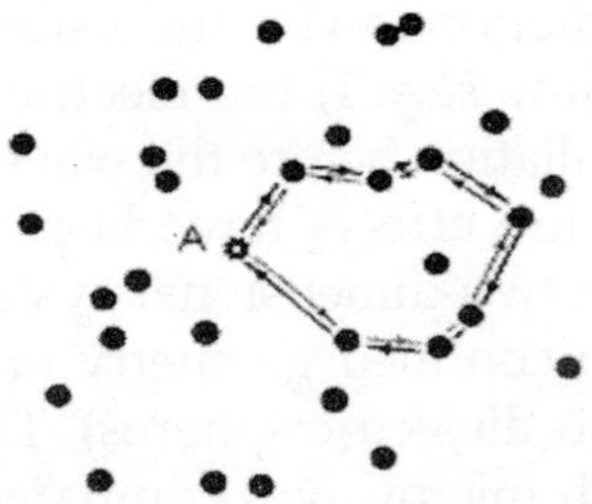

Figure 4.1: Anderson localization of waves in disordered systems originates from interference in multiple elastic scattering.

Here we consider a light source (like an excited atom emitting a photon) in a disordered medium at position A. The light source is denoted by a star symbol and the spheres denote the scattering elements. A random light path that returns to the light source can be followed in two opposite directions. The two waves which propagate in opposite directions along this loop will acquire the same phase and therefore interfere constructively in A. This leads to a higher probability of the wave coming back to A and consequently a lower probability of propagating away from A. On decreasing the mean free path *l*, the probability for such looped paths increases and at strong enough scattering the system makes a phase transition from the normal conducting state into a localized state, due to interference. In the localized regime, the system behaves as a non-absorbing insulator. Light which is incident on, for example, a slab would be almost completely reflected and the remaining transmission would decrease exponentially with the slab thickness.

The main difficulty in the search for localization of light has been the realization of strong enough scattering. The appropriate measure for the amount of scattering is the mean free path *l* for the light in the medium, times the magnitude of the wavevector *k*. Localization is expected for $kl \leqslant 1$, which is known as the (modified) Ioffe–Regel criterion. This criterion can be understood intuitively if one realizes that below $kl = 1$, the electric field cannot even perform one oscillation before the wave is scattered again. So far, localization effects have been reported only for microwaves in a two-dimensional system of rods and for microwaves in a confined geometry (a copper tube filled with metallic and dielectric spheres). The disadvantage of experiments with microwaves compared to light waves is that it is difficult to avoid absorption.

It is found that semi-conductor powders acts as very strongly scattering samples for light waves. Semi-conductors can have a very large refractive index, while, for wavelengths in the bandgap, the absorption is extremely small.

To characterize the mean free path l, coherent backscattering can be used. This phenomenon is a general interference effect between counter-propagating waves, which leads to a narrow cone in exact backscattering. It is seen as the precursor to Anderson localization and is therefore also called 'weak localization of light'. The width of the backscattering cone and the angle of its cusp are inversely proportional to l and therefore enable the determination of the amount of scattering inside the samples. Furthermore, the cusp is due to a summation up to (in principle) infinite path length, and is therefore only present in the absence of absorption.

In a transmission coefficient study samples with different particle diameters are illuminated with a mode-locked Nd:YAG laser. The transmission coefficient is defined as the ratio between total transmitted flux and the incident flux. This quadratic dependence of the transmission coefficient on the inverse thickness is the expected behavior at an Anderson localization transition.

For classical diffusion transmission coefficient (T) would decrease linearly with the sample thickness L. It is found that for these samples T shows a quadratic dependence

($T \propto L^{-2}$). This is exactly the behavior predicted by the scaling theory of localization at the localization transition.

If we decrease the particle size even further we expect (from Mie-scattering theory) to obtain even stronger scattering. The transport of light has come to a halt owing to interference, and the system has made a phase transition from a conducting into a localized state. The characteristic length scale for the exponential decay is called the localization length.

4.2 Localization of Superradiance near a PBG

When the atomic resonance frequency lies at the band edge, superradiant emission remains localized in the vicinity of the atoms. This leads to a steady state with spontaneously broken symmetry in which the atomic system acquires a macroscopic polarization. In strongly disordered dielectrics have revealed signatures of incipient localization [7, 8], the proposal of creating a complete PBG [9,10], offers the most systematic route to this goal. In recent years, several dielectric structures have been predicted [11-13] and observed [13] to exhibit a PBG, a range of frequencies for which no propagating electromagnetic modes are allowed. The existence of PBG materials gives rise to a number of interesting phenomena including the suppression of spontaneous emission [9], the formation of strongly localized states of light,[10] and photon-atom bound states [14]. Spontaneous emission in a PBG displays distinct features from those in free space such as oscillatory behavior, fractional steady-state atomic population on the excited state, and subnatural line width [15, 16]. These are all direct consequences of localization. The strong dependence of the collective processes on the structure of the PBG can be associated with the density of states near the band edge of a PBG. Although PBGs are analogous to electronic bandgaps in semi-conductors, there are many intriguing aspects of photons, which are not shared by electronic systems. Among these are laser action and superradiance. These are related to the bosonic nature of light through which many photons can occupy the same mode. The recent observation of laser action in strongly scattering media motivates studies in this new direction.

This novel form of spontaneous symmetry breaking is the analog of lasing without a cavity mode. The collective emission near the photonic band edge is accompanied by self-induced oscillations, a simple illustration of the ringing regime in superradiance. In addition to being a fundamental phenomenon, localization of superradiance may play an important role in low threshold microlasers based on PBG engineering. It suggests that a light emitting diode operating near a photonic band edge will exhibit very high modulation speed and coherence properties without recourse to external mirrors or even a true cavity mode.

In conclusion, the localization and macroscopic coherence in superradiant emission occur near a photonic band edge even in the absence of a dielectric defect mode or other cavity mode. This suggests the possibility that ordinary light emission in a perfectly periodic dielectric may exhibit coherence properties with infinitesimal threshold. For the case of a physical three-dimensional anisotropic gap, the superradiant emission occurs much faster and with higher peak intensity than conventional superradiance. These results are based on a simple model of point superradiance in which N atoms are confined to a region smaller than the wavelength of light. The spontaneous atomic polarization in the steady state is analogous to the emergence of a superfluid order parameter for photons. If the volume of superradiance is made larger than a cubic wavelength, alternate forms of spontaneous symmetry breaking may arise from the spatial dependence of the effective atomic resonance dipole-dipole interaction (RDDI). In general, RDDI causes a breaking of the permutation symmetry of the *N*-atom wave function [10]. This may diminish the magnitude of the macroscopic steady-state polarization, and lead to a "Bose glass" under certain circumstances. Following the initial prediction of this effect, attention has focused on systematic methods for the experimental realization of strong localization of light and its consequences in laser physics.

4.3 Coherent Control of Spontaneous Emission near a Photonic Band Edge

The coherent control of spontaneous emission for a three-level atom located within a photonic band gap (PBG) material, with one resonant frequency near the edge of the PBG. Spontaneous emission from the three-level atom can be totally suppressed or strongly enhanced depending on the relative phase between the steady-state control laser coupling the two upper levels and the pump laser pulse used to create an excited state of the atom in the form of a coherent superposition of the two upper levels. Unlike the free-space case, the steady-state inversion of the atomic system is strongly dependent on the externally prescribed initial conditions. This non-zero steady-state population is achieved by virtue of the localization of light in the vicinity of the emitting atom. It is robust to decoherence effects provided that the Rabi frequency of the control laser field exceeds the rate of dephasing interactions. As a result, such a system may be relevant for a single-atom, phase-sensitive optical memory device on the atomic scale. The protected electric dipole within the PBG provides a basis for a qubit to encode information for quantum computations [46, 47].

In addition to strong localization of light [2] at the classical level, this leads to the suppression of spontaneous emission [1, 17] and the formation of photon atom bound states [18]. Near a photonic band edge, spontaneous emission dynamics is anomalous [19] and leads to a fractionalized steady-state inversion for a single atom. In a PBG system the emitted photon remains partially localized in the vicinity of the emitting atom. This, in turn, leads to long-term memory effects and non-Markovian behavior in the collective spontaneous emission from many atoms. These remarkable changes in the radiative properties of atoms and molecules arise purely from the dielectric environment of the PBG host, without the presence of an external field [20]. On the other hand, coherent interaction of atoms with external laser fields can have a profound effect on radiative dynamics [20], even in ordinary vacuum [21-25]. In this system, a pump laser pulse is used to create an excited state of the atom with an atomic Bloch vector specified by the area of the incident pulse. A control laser field with a specific phase relation to the pump laser pulse stimulates radiative transitions between the upper two excited states. It is shown that spontaneous emission can be totally suppressed or strongly enhanced by changing the optical paths and, hence, the relative phase, between the control laser field and the initial atomic Bloch vector determined by the pump laser pulse. Unlike the free space case, the steady-state inversion of the atomic system is strongly dependent on the externally prescribed initial conditions. As a result, such a system may be relevant for a single atom, phase-sensitive, optical memory device.

In conclusion, we have considered coherent control of spontaneous emission near the edge of a PBG. Because of quantum interference, as well as coherent photon localization, spontaneous emission can be totally suppressed or strongly enhanced, depending on the relative phase of the initial atomic Bloch vector (determined by the pump laser pulse) and the control laser field. The steady-state atomic population on the upper levels keeps the memory of the atomic initial state as well as the phase of the external field.

4.4 Multiphoton Localization

The many-particle spectrum of an isotropic frequency gap medium doped with impurity resonance atoms is studied using the Bethe ansatz technique. The spectrum is shown to contain pairs of quantum correlated "gap excitations" and their heavy bound complexes ("gap solitons"), enabling the propagation of quantum information within the classically forbidden gap. In addition, multiparticle localization of the radiation and the medium polarization occurs when such a gap soliton is pinned to the impurity atom.

Light localization is a classical effect predicted to occur in strongly scattering dielectric microstructures. In the context of PBG materials, non-classical forms of localization such as photon-atom bound states have been predicted when the resonant transition frequency of an impurity atom lies within a gap. This bound state is an eigenstate of the quantum electrodynamic Hamiltonian for a realistic PBG crystal exhibiting a general anisotropic photon dispersion relation. In this state, a virtually emitted photon may tunnel many wavelengths away from the atom before being reabsorbed, leading to non-Markov memory effects in collective light emission from many atoms. It was recently shown that an effective model describing both isotropic PBG systems and frequency dispersive media (DM) doped with resonance atoms exhibits hidden integrability and are diagonalized exactly by means of the Bethe ansatz technique. This suggests the possibility of a rich multiparticle spectrum in real physical systems exhibiting a frequency gap, when such systems are doped with impurity atoms.

In addition to ordinary polaritons and their bound complexes (ordinary solitons) occurring outside of the gap, the subgap spectrum of the system is shown to contain propagating pairs of correlated "gap excitations" and their heavy bound complexes (gap solitons). The individual gap excitations comprising the pair are correlated such that the probability amplitude of finding them far apart decreases exponentially with the ratio of their separation distance to the classical penetration length of the radiation into the medium. In addition to heavy gap solitons propagating within the gap, the spectrum contains multiphoton localized states pinned to the atom. Under external perturbations a pinned gap soliton may dissociate into propagating gap excitations. We evaluate the dispersion relations, the effective masses, and the dissociation energies of quantum gap solitons and we show that they are stable with respect to weak perturbations. Our results demonstrate a clear distinction, to the quantum level, between fermionic gap systems (such as electronic semi-conductors) and bosonic gap systems. In a semi-conductor, propagation within the energy gap is strictly forbidden. In the bosonic gap, however, certain nonclassical many-body gap states are allowed to propagate. This may have important consequences for the transmission of quantum information within a bosonic gap medium. Although the Bethe ansatz method made use of the isotropic one-particle dispersion relation, qualitative similar results may hold in a realistic, anisotropic PBG material, which also exhibits a small non-resonant Kerr non-linearity.

4.5 Photon Localization Barrier

While quantum electrodynamics (QED) underwent an impressive development and reached its maturity in the middle of the last century, one of its basic concepts - the photon wave function in free space - was deprived of such fortune. Although the photon wave function in coordinate representation was introduced already in 1930 by Landau and Peierls [26] the concept was found to suffer from inherent difficulties that were not overcome during the century. The common explanation may be summed up as follows:

(1). no position operator exists for the photon,

(2). while the position wave function may be localized near a space-time point, the measurable quantities like the electromagnetic field vectors, energy, and the photo detection probability remains spread out due to their non-local relation with the position wave function.

However, just before the turn of the century both of these widely-espoused notions were disproved [27, 28] and in the new century a fresh interest in the photon localization problem seems to have been awakened, meeting the needs of developments in near-field optics, cavity QED, and quantum computing. Bialynicki-Birula writes [28] that the statement "even when the position wave function is strongly concentrated near the origin, the energy wave function is spread out over space asymptotically like $r^{7/2}$" is incorrect and that both wave functions may be strongly concentrated near the origin. He demonstrates, on one hand, that photons can be essentially better localized in space - with an exponential falloff of the photon energy density and the photo detection rates. On the other hand, he establishes - and it is even some what startling that nobody has done it earlier - that certain localization restrictions arise out of a mathematical property of the positive frequency solutions which therefore are of a universal character and apply not only to photon states but hold for all particles.

4.6 Photon Localization Laser

Lasing is facilitated by resonant excitation of localized modes at the pump laser wavelength, which are peaked deep within the sample with greatly enhanced intensity. Emission occurs into long-lived localized modes overlapping the localized gain region. This mechanism overcomes a fundamental barrier to reducing lasing thresholds in diffusive random lasers, in which multiple scattering restricts the excitation region to the proximity of the sample surface. In conventional lasers, emission is stimulated into well-defined cavity modes and emerges as a coherent beam. Care is taken to suppress scattering within the cavity since this would shorten the photon residence time in the lasing mode and thereby raise the excitation power required to initiate lasing. In the opposite limit of a random amplifying medium, however, multiple scattering impedes the flow of light out of the gain region [29, 30]. This has led to a search for lasing action in random amplifying media, which was observed in rare earth [31] and semi-conductor [32] powders, colloidal suspensions in dye solutions [33, 34], and organic films [35]. However, the lasing threshold in these diffusive systems is not appreciably lowered below that for amplified spontaneous emission in homogeneous media because multiple scattering also impedes the flow of the incident pump light into the sample [36]. This creates a shallow gain region and allows subsequently emitted photons to escape promptly as diffuse luminescence.

A collimated laser beam is produced at greatly reduced lasing threshold in a stack of microscope cover slides with interspersed dye films. The threshold is sufficiently low for lasing to be induced with a quasi-cw illumination by a 3-Watt laser beam from an Argon-ion laser. Pump energy is deposited deep within the sample, where it is greatly enhanced, by the resonant excitation of spectrally narrow localized modes. Subsequent emission occurs into long-lived localized modes, which spatially overlap the region of molecular excitation. This is demonstrated by the strong correlation of pump transmission and output laser power.

The sample is a stack of L partially reflecting glass slides of random thickness of ~100 μm with an amplifying region or an air gap between the slides. The amplifying layers consisted either of 25-μm thick solution of Rhodamine 6G laser dye in ethylene glycol, or of a single 0.3-mm-thick chromophore-doped plastic sheet. The stack was illuminated at normal incidence by a 150 ns pulse from a frequency-doubled, Q-switched Nd:YAG laser at 532 nm focused to a 100-μm spot. Light in this one-dimensional (1D) geometry is localized by multiple scattering from the parallel layers, which returns the wave upon itself [37-40]. Emission spectra, recorded with a 0.07-nm-resolution spectrometer, are shown in Figure 4.1.

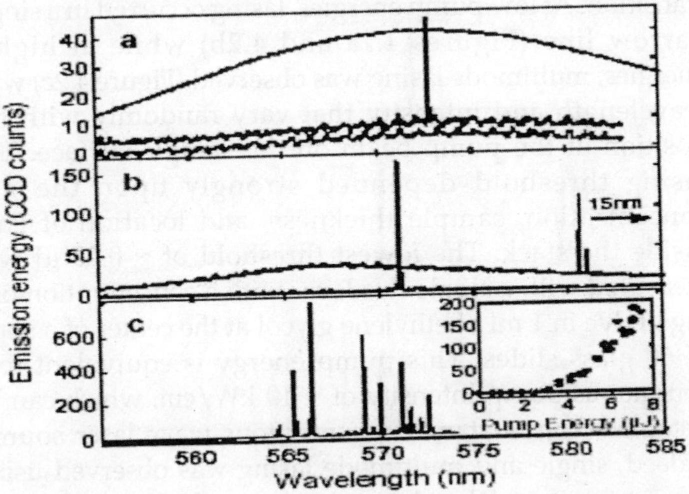

Figure. 4.2: Emission spectra of the neat dye solution (dashed curve in (a)) and of localization laser for the following experimental parameters: (a) - single 25-μm dye layer (2 mg of dye in 1 ml of ethylene glycol) in the middle of a stack of 50 glass slides; pump energy, *E* pump, was 110 μJ and 130 μJ per pulse for the lower and middle curves (shifted up for clarity), respectively. (b, c) - sample of 100 layers, in which every other air gap was filled with the same dye solution; *E* pump = 3.2 μJ and 4.9 μJ per pulse for (b) and (c), respectively. The inset in (c) shows the total emission energy integrated over the spectral region 550-590 nm. Dashed curve in (b) shows the emission spectrum (shifted by 15 nm) obtained with a single colored plastic sheet in the center of a stack of 50 glass slides, which was pumped with a chopped 3-Watt cw beam from an Argon ion laser.

The broad emission spectrum of neat dye solution (dashed curve in Figure 4.2a) is compared with the emission spectrum from the random stack with interspersed dye layers slightly below and above the lasing threshold. Below threshold, the spectrum shows resolution-limited peaks of the electro-magnetic modes of the system (lower curve in Figure 4.2a). Above threshold, a collimated emission beam perpendicular to the sample layers was observed (middle curve in Figure 4.2a). The abrupt change in the output power with increasing pump energy shown in the inset to Figure.4.2 c is characteristic of the lasing transition. At low pump energies, lasing occurred in a single narrow line (Figures 4.2a and 4.2b) while at higher energies, multimode lasing was observed (Figure 4.2c) with wavelength and intensity that vary randomly with the position of the pump beam on the sample surface. The lasing threshold depended strongly upon the dye concentration, sample thickness, and location of gain inside the stack. The lowest threshold of ~ 0.25 μJ was measured with a single dye layer with a concentration of 4 mg of dye in 1 ml of ethylene glycol at the center of a stack of 60 glass slides. This pump energy is equivalent to a continuous pump intensity of ~ 10 kW/cm, which can be obtained from a standard continuous-wave laser source. Indeed, single and multimode lasing was observed using the output of a 6-Watt Argon-ion laser which was chopped in order not to overheat the sample (dashed curve in Figure. 4.2 b). The distinct spontaneous emission and lasing peaks, shown in Figure 4.2, contrast with the continuous emission spectrum of the diffusive random laser and is characteristic of localized electromagnetic modes.

In localized media, the spectral width of a mode, which is the inverse of the escape time of photons from the medium, is smaller than the typical spacing between modes In the presence of substantial gain, emission into a few long-lived modes of the microscopic sample is enhanced, giving rise to distinct spectral peaks [41]. In contrast, the overlap of many modes in macroscopic diffusive lasers results in non-resonant feedback and a continuous emission spectrum [30]. Because of the large number of competing modes within the excitation volume, the number of stimulated photons in each mode is low. As a result, the lasing threshold is high and only a modest spectral narrowing is achieved. Even though the average optical transport is suppressed, resonant tunneling through localized states provides spectrally narrow modes, in which transmission can be high [40, 42, 43]. The laser beam was focused to an 80-μm-diameter spot. Stronger focusing was avoided in order to prevent spectral line broadening due to the angular spread of the incident beam. Since lasing occurs via the excitation of narrow localized modes, large fluctuations in the output lasing power may be expected as different parts of the sample are excited [44]. In the present case, strong transmission is expected when the pump laser is at resonance with localized modes, which are spatially peaked near the center of the sample. Pump intensity is therefore exponentially enhanced within the sample, which results in efficient energy transfer to the gain medium. Further, since excitation in the center of the sample is likely to escape via emission into the localized modes with long photon residence time, the opportunity for stimulated emission is enhanced and the laser output power is high.

4.7 Photon Localization in Resonant Media

Photon localization has been of intense interest since it was first proposed as the analog of electron localization [45]. Accurate measurements of the statistics of the electromagnetic field and intensity and of the total transmission in random samples have deepened our understanding of mesoscopic physics and localization. However, the very possibility of photon localization in three-dimensional systems without substantial crystalline order has been called into question by the barriers to achieving and detecting electromagnetic localization. Unlike electrons that can be trapped by the Coulomb interaction at atomic sites, photons are not bound by individual particles. They are not strongly scattered by particles either, except at Mie resonances where the scattering cross-section can considerably exceed the geometric crosssection. To preserve the strength of scattering, however, particles must be separated by at least the diameter of the scattering crosssection, requiring particle separation that is larger than the wavelength λ. It would then appear that the Ioffe-Regel-like condition for localization in three dimensions [46] $l_{sc} << \lambda/2\pi$ where l_{sc} is the scattering mean free path, cannot be satisfied by collections of independent resonant scatterers [47].

Measurements of exponential scaling of transmission [48-50] have not definitively established localization because the possible presence of absorption could also lead to exponential decay of intensity within a sample. However, recent measurements of coherent backscattering in macro porous GaP networks [52] along with theoretical predictions suggest that the approach to localization can be observed in the rounded backscattering peak from weakly absorbing samples. Also recently, the variance of relative fluctuations has been shown to provide a decisive test for localization, even in the presence of strong absorption [53]. This provides a sure guide in the search for localization and raises the possibility that measurements of key scattering parameters can be used to sort out the precise material and structural characteristics that may edge samples towards and potentially across the localization threshold. Localization does not occur on resonance because the narrowing of the level width is offset by the enhancement of the density of states.

4.8 Electromagnetic Field Enhancement in Laser-Irradiated, Random Porous Media

Radiative transport in random porous media has long been an important problem in science and engineering. As a laser beam is irradiated on a non-absorbing random porous medium in which the particle size is much larger than the laser wavelength, the transmittivity reveals a diffusive behavior; i.e., decays inversely with the medium thickness. Recently, however, it has been found that as the particle size decreases to the order of the laser wavelength, the diffusion breaks down and the transmission coefficient decays exponentially with the medium thickness, due to the constructive interference among the multiply scattered waves. This phenomenon, termed "photon localization," is the counterpart of electron localization suggested by Anderson and has been investigated theoretically and experimentally in one-, two-, and three-dimensional structures. With the onset of localization, the electromagnetic (EM) waves may not propagate through, but are confined in a finite spatial region, forming a random resonator.

4.9 References

1. E. Yablonovitch, *Phys. Rev. Lett.* **58**, 2059 (1987).
2. S. John, *Phys. Rev. Lett.* 58, 2486 (1987).
3. K. M. Ho, T. J. Chan, and C. M. Soukoulis, *Phys. Rev.Lett.* 65, 3152 (1990).
4. S. Fan, P. Villeneuve, R. Meade, and J. Joannopoulos,*Appl. Phys. Lett.* 65, 1466 (1994).
5. E. Yablonovitch, T. J. Gmitter, and K. M. Leung, *Phys.Rev. Lett.* 67, 2295 (1991). For a review of recent progress in the microfabrication of PBG materials, see *Photonic Band Gap Materials,* edited by C. M. Soukoulis, NATO ASI Ser. E Vol. 315 (Kluwer Academic, Dordrecht,1996).

6. Diederik S et al, *Nature* **390**, 671-673 (18 December 1997).
7. N. Garcia and A. Z. Genack, *Phys. Rev. Lett.* **66**, 1850(1991).
8. A. Z. Genack and N. Garcia, *Phys. Rev. Lett.* **66**, 2064(1991).
9. E. Yablonovitch, *Phys. Rev. Lett.* **58**, 2059 (1987).
10. Anderson, P. W. *Phil. Mag.* B 52, 505–09 (1985).
11. K. M. Ho, C. T. Chan, and C. M. Soukoulis, *Phys. Rev.Lett.* **65**, 3152 (1990).
12. Z. Zhang and S. Satpathy, *Phys. Rev. Lett.* **65**, 2650(1990).
13. E. Yablonovitch, T. J. Gmitter, and K. M. Leung, *Phys.Rev.Lett.* **67**, 2295 (1991).
14. S. John and J. Wang, *Phys. Rev. Lett.* **64**, 2418 (1990); *Phys. Rev.* B **43**, 12 772 (1991).
15. R. F. Nabiev, P. Yeh, and J. J. Sanchez-Mondragon, *Phys.Rev.* A **47**, 3380 (1993).
16. S. John and T. Quang, *Phys. Rev.* A **50**, 1764 (1994).
17. R. F. Nabiev, P. Yeh, and J. J. Sanchez-Mondragon, *Phys.Rev.* A **47**, 3380 (1993).
18. S. John and J. Wang, *Phys. Rev. Lett.* **64**, 2418 (1990);Phys. Rev. B **43**, 12 772 (1991); S. John, in "Confined Eletrons and Photons", edited by E. Burstein and C. Weisbuch, *Localization of Light in Disordered and Periodic Dielectrics* (Plenum, New York, 1995).
19. S. John and Tran Quang, *Phys. Rev.* A **50**, 1764 (1994).
20. M. Lewenstein and T. W. Mossberg, *Phys. Rev.* A **38**,808(1988); G. rochmalicki and M. Lewenstein, *Phys. Rev.*A **40**, 2517 (1989).
21. M. O. Scully, S. Y. Zhu, and A. Gavrilides, *Phys. Rev.Lett.* **62**, 2813 (1989); A. Imamoglu, *Phys. Rev.* A **40**,2835 (1989).
22. L. M. Narducci, M. O. Scully,G. L. Oppo, P. Ru, and J. R.Tredice, *Phys. Rev.* A **42**, 1630 (1990); G. S. Argarwal,*Phys. Rev.* A **54**, 3734 (1996).
23. G. S. Agarwal, *Phys. Rev. Lett.* **67**, 980 (1991); S. Y. Zhu and M.O. Scully, *Phys. Rev. Lett.* **76**, 388 (1996).
24. P. Brumer and M. Shapiro, *Accts. Chem. Res.* **22**, 407(1989); C. Chen, Y-Y. Yin, and D. S. Elliot, *Phys. Rev.Lett.* **64**, 507 (1990).

25. For a more detailed review, see *Coherence Phenomenain Atoms and Molecules in Laser Field,* edited by A. D. Bandrauk and S. C. Wallace (Plenum, New York, 1992).

26. L. D. Landau and R. Peierls, *Z. Phys.* 62, 188 (1930).

27. M. Hawton, *Phys. Rev.* A, 59, 3223 (1999).

28. I. Bialynicki-Birula, *Phys. Rev. Lett.* 80, 5247 (1998).

29. V. S. Letokhov, Sov. *Phys. JETP* **26**, 835 (1968).

30. R. V. Ambartrsumyan, N. G. Basov, P. G. Kryukov, and V. S. Letokhov, in *Progress in Quantum Electronics*,edited by J. H. Sanders and K. W. H. Stevens (Pergamon, Oxford, 1970), p. 107.

31. C. Gouedard, D. Husson, C. Sauteret, F. Auzel, and A.Migus, *J.Opt. Soc. Am.* B **10**, 2358 (1993); V. M. Markushev, V. F. Zolin, and C. M. Briskina, *J. Quantum Electron.* **16**, 281 (1986); G. R. Williams, S. B. Bayram,S. C. Rand, T. Hinklin, and R. M. Laine, *Phys. Rev.* A **65**, 013807 (2002).

32. H. Cao, Y. G. Zhao, S. T. Ho, E. W. Seelig, Q. H. Wang, and R.P.H. Chang, *Phys. Rev. Lett.* **82**, 2278 (1999).

33. N. M. Lawandy, R. M. Balachandran, A. S. L. Gomes,and E. Sauvain, *Nature* **368**, 436 (1994).

34. W. L. Sha, C.-H. Liu, and R. R. Alfano, *Opt. Lett.* **19**,1922 (1994); M. A. Noginov, H. J. Caulfield, N. E.Noginova, and P. Venkateswarlu, *Opt. Comm.* **118**, 430(1995).

35. S. V. Frolov, Z. V. Vardeny, K. Yoshino, A. Zakhidov, and R. H. Baughman, *Phys. Rev.* B **59**, R5284 (1999).

36. A. Z. Genack and J. M. Drake, *Nature* **368**, 400 (1994);D. Wiersma, M. P. v. Albada, and A. Lagendijk, *Nature* **373**, 203 (1995).

37. P. W. Anderson, *Phys. Rev.* **109**, 1492 (1958); M. V.Berry and S.Klein, *Eur. J. Phys.* **18**, 222 (1997).

38 . D. J. Thouless, *Phys. Rev. Lett.* **39**, 1167 (1977).

39. G. C. Papanicolaou, *SIAM J. Appl. Math.* 21, 13 (1971); P. Sheng, B. White, Z. Zhang, and G. Papanicolaou,*Phys. Rev.* B **34**, 4757 (1986).

40. M. Y. Azbel and P. Soven, *Phys. Rev.* B **27**, 831 (1983).

41. V. M. Apalkov, M. E. Raikh, and B. Shapiro, *Phys. Rev.Lett.* **89**,016802 (2002).

42. D. J. Thouless, *Phys. Rev. Lett.* **39**, 1167 (1977).

43 . V. Freilikher, M. Pustilnik, and I. Yurkevich, *Phys. Rev.*B **56**, 5974 (1997); X. Jiang and C. M. Soukoulis, *Phys.Rev. Lett*. **85**, 70(2000); P. Pradhan and N. Kumar,*Phys. Rev.* B **50**, 9644 (1994); C. Vanneste and P. Sebbah, *Phys. Rev. Lett.* **87**, 183903 (2001); Z. Q. Zhang, *Phys. Rev.* B **52**, 7960 (1995).

44. A. L. Burin, M. A. Ratner, H. Cao, and S. H. Chang, *Phys. Rev. Lett.* **88**, 093904 (2002).

45. S. John, *Phys. Rev. Lett.* 53, 2169 (1984).

46. A. Lagendijk and B. A. van Tiggelen, *Phys. Rep.* **270**, 143(1996).

47. A. Lagendijk, in *Current Trends in Optics,* edited by J. C. Dainty (Academic, London, 1994).

48. A. Z. Genack and N. Garcia, *Phys. Rev. Lett.* **66**, 2064(1991).

49. D. S. Wiersma, P. Bartolini, A. Lagendijk, and R. Righini, *Nature* (London) **390**, 671 (1997); D. S. Wiersma, J. G.Rivas, P.Bartolini, A. Lagendijk, and R. Righini, *Nature* (London). **398**, 207 (1999).

50. Yu.A. Vlasov, M. A. Kaliteevski, and V. V. Nikolaev, *Phys.Rev.* B **60**, 1555 (1999).

51. F. J. P. Schuurmans, M. Megens, D. Vanmaekelbergh, and A. Lagendijk, *Phys. Rev. Lett.* **83**, 2183 (1999).

52. B. A. Van Tiggelen, A. Lagendijk, and D. S.Wiersma, *Phys.Rev. Lett.* **84**, 4333 (2000).

53. A. A. Chabanov, M. Stoytchev, and A. Z. Genack, *Nature* (London) **404**, 850 (2000).

CHAPTER 5

APPLICATION OF PHOTONIC BANDGAP MATERIALS

Chapter content

5.1 Introduction

Photonic-bandgap materials are in one of the fastest growing subfields in optical physics. Photonic-bandgap (PBG) crystals have immense promise for manipulating and engineering the local photonic densities of states (DOS) and realizing novel electromagnetic phenomena. By enhancing or suppressing the local photonic DOS, spontaneous emission can be tailored, creating novel phenomena in optical physics. Researchers suggests that PBG materials exhibit fundamentally new physics such as photon-atom bound states, lasing without a cavity mode, quantum optical spin-glass states of impurity atomic dipoles, and optical gap solitons. Applications include zero-threshold micro-lasers with high modulation speed and low threshold optical switches and all-optical transistors for optical telecommunications and high speed optical computers.

Photonic crystals offer the possibility of controlling and manipulating light within a given frequency range through the presence of photonic bandgap. The small size of photonic bandgap structures makes it possible to fabricate ultrasmall optical devices like lasers and cavities. These cavities find applications in several novel fields including cavity quantum electrodynamics (QED) and quantum computations. The ability to control the propagation of electromagnetic waves within the photonic bandgap (by including appropriate defects to the photonic crystal structure) makes the photonic bandgap materials very attractive for the fabrication of all-optical integrated circuits. Furthermore, photonic crystal structures provide the ability for custom designing the electromagnetic mode patterns, frequencies and numbers. This opens up the possibility of novel devices (miniature lasers and detectors at different frequencies, switches, filters, interconnects, etc.) and interesting physics. The outstanding potentials of photonic bandgap devices to revolutionalize communications, information, and energy technologies along with the existence of several challenges in design, optimization, fabrication, and characterization have inspired extensive research activities in the field of photonic bandgap materials and devices. The main applications are:

- Higher efficiency antennas and receivers at microwave and millimeter wave frequencies
- High Q,frequency selective filters
- PBG structures at near IR and optical wavelengths
- PBGs for photocatalysis
- Effects of defects.

Since the invention of the laser, the field of photonics has progressed through the development of engineered materials, which mold the flow of light. Applications include zero-threshold micro-lasers with high modulation speed and low threshold optical switches and all-optical transistors for optical telecommunications and high-speed optical computers. Applications such as telecommunications, data transfer, and computing will be greatly enhanced through all-optical processing in which bits of information, encoded in the form of a photon number distribution, can be transmitted and processed without conversion to and from electrical signals. In this manner, the information contained in an entire encyclopedia can be transmitted over a fibre optic phone line in a matter of seconds. This relies on the development of ultra-low noise coherent light sources. We are evaluating the quantum statistics of photons produced by laser emission in a PBG material to evaluate the extent of signal noise reduction referred to as photon-antibunching and squeezing. These are crucial to lowering the bit error rate in optical telecommunication networks. In a PBG material the drastic reduction of spontaneous emission as well as the reduction of propagative pathways for photons, facilitates the realization of very low quantum noise.

5.2 Quantum Computing

The PBG material provides dopant atoms with a high degree of protection from damping effects of spontaneous emission and dipole dephasing. In this case the two-level atom may act as a two-level quantum mechanical register or single photon logic gate for all optical quantum computing. There are two models for quantum computation within PBG materials. The first involves an atom, which is laser-cooled in the void regions of a PBG material. In this case a polarized photon (flying qubit) with frequency just outside of the gap excites a protected atomic level inside the gap (stationary qubit) by resonant coupling to a third atomic level just outside the gap. The second and third atomic levels are coupled by an external laser field, which drives a two-photon transition. This single atom acts as a phase sensitive quantum memory device. The resulting qubit is robust to decoherence effects provided that the Rabi frequency of the coherent laser field exceeds the rate of dephasing interactions. In effect, coherence is externally imposed on the system. In the second model for quantum computing, the single photon occupation of a localized field mode within an engineered network of defects in a PBG material acts as the qubit. Qubit operations are then mediated by optically excited atoms interacting with these localized states of light as the atom traverses an open channel within the extensive void network of the PBG material. Here, quantum information is stored in localized states of light. Quantum computers can perform calculations in relatively short polynomial time. These calculations require prohibitively long exponential time with any classical computer. PBG materials may provide the essential hardware for the realization of a quantum computer. For a quantum computer to function, quantum bits of information must be stored as well as transferred from point to point without decoherence effects taking place. The PBG is an ideal environment for protecting photons and atoms from decoherence over extended periods of time.

5.3 Exceptionally Directional Sources with PBG Crystals

Three-dimensional photonic-bandgap crystals are used to design and fabricate uniquely directional sources and receivers. By utilizing the resonances of a Fabry–Perot cavity formed with photonic-bandgap crystals, we were able to create exceptionally directional sources by placing the sources within such a cavity. The early application of photonic crystals to the field of antennas and emission was to use the photonic crystals as a perfect reflector at frequencies within the bandgap and to place sources on the surface of the photonic crystal. When such antennas were driven at frequencies within the bandgap all the power was radiated in the air side with no power emerging from the PBG side [3, 4, 5]. This substantially reduced the large power loss of antennas on semiconductor substrates, where most of the power (>90per cent) can be lost as substrate modes. However, the antennas on PBG substrates had radiation patterns that were sensitive to the height and lateral position of the antenna. This is because the antenna power depends on the local electric field at the substrate, which is a sharply varying function of the lateral position and height of the antenna on the substrate [3, 5]. The radiation patterns of the present resonant antenna are far superior to those achieved with antennas on the PBG crystals. We also find resonant antennas are less sensitive to position within the cavity. However, the bandwidth of resonant antennas is less than for the antennas on PBG substrates. Some preliminary experiments with atoms in optical cavities have already shown a narrowing of the emission pattern at the resonant frequency. Optical measurements have been taken that find sidelobes and a broadening of the emission pattern away from resonance, in good agreement with our simulations. The resonant cavities offer a practical way to test predictions expected in optical physics, with the antenna being the analog of an emitting atom with cavity dimensions on the nanoscale, which can lead to a fundamental understanding of cavity electrodynamics. We anticipate that uniquely directional high-gain sources and detectors can be generated with this method.

5.4 High-Q Microwave Cavities Based on PBG Structures

Photonic bandgap (PBG) structures have gained increasing recognition due to their application potential ranging from the microwave - up to the visible light frequency range. A novel approach of high-Q resonators for microwave frequencies based on extended defects is investigated in two-dimensional photonic bandgap structures machined from high-purity aluminium oxide. Numerical field simulations and experimental investigations indicate that a hexagon shaped air-cavity arranged inside a two-dimensional hexagonal lattice of holes in a dielectric wafer supports a quasi transverse-electric (TE) mode with six fold symmetry of the electromagnetic field energy distribution. For this mode about 60 per cent of the electromagnetic field energy is stored inside the air cavity leading to a significant reduction of dielectric losses in comparison to classical dielectric resonators. Field confinement in the direction perpendicular to the photonic bandgap structure can be achieved by employing a stack of dielectric wafers with PBG structures inside a metallic shielding cavity.

5.5 3-D Photonic Bandgap Structures in the Microwave Regime

Three-dimensional photonic bandgap (PBG) structures using alumina (Al_2O_3) as the high permittivity material were modeled and then the structures were fabricated by Fused Deposition of Multi-materials (FDMM) technology (Figure 5.1). A finite element method and real-time electromagnetic wave propagation software were used to simulate and design the layered PBG structures for applications in the microwave frequency range. The modeling predicted a 3-D photonic bandgap in the 16.5–23.5 GHz range. FDMM provides a computer-controlled process to generate 3-D structures, allowing high fabrication flexibility and efficiency. Electromagnetic measurements displayed the presence of a bandgap between 17.1–23.3 GHz, showing a good agreement with the predicted values. These PBG structures are potential candidates for applications in advanced communication systems.

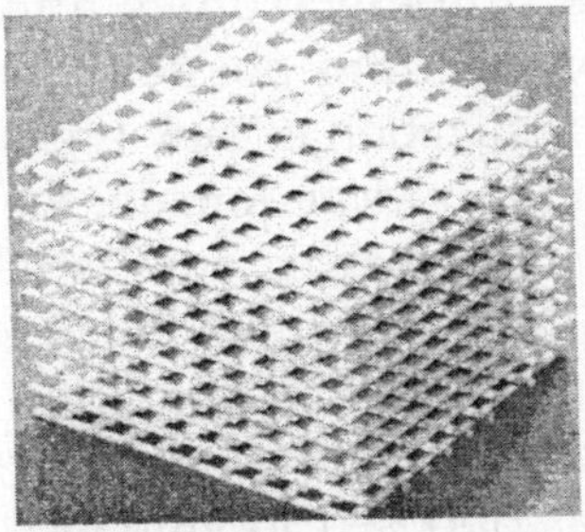

Figure 5.1: PBG structure obtained by stacking alumina rods. The gap is between 12 and 14 GHz

5.6 Active Photonic Bandgap Structures

One-, two-, and three-dimensional photonic bandgap structures have been the topic of many investigations because their properties can be engineered to achieve useful functionalities. Very recently, PBG structures with tunable optical properties have been proposed and demonstrated. Such active PBG structures could find applications in many technological areas, such as optical interconnects and biosensors.

5.7 Waveguidance by PBG Effect in Optical Fibers

Photonic crystals form a new class of intriguing building blocks to be utilized in future optoelectronics and electromagnetics. One of the most exciting possibilities offered by photonic crystals is the realization of new types of electromagnetic waveguides. In the optical domain, the most mature technology for such (PBG) waveguides is in optical fibre configurations. These new fibres can be classified in a fundamentally different way to all optical waveguides and possess radically different guiding properties due to PBG guidance, as opposed to guidance by total internal reflection.

Since the first proposals of a (PBG) effect by Yablonovitch and John [6,7], this subject has received a rapidly increasing research interest [8]. The fact that periodic dielectric structures may be designed to exhibit electromagnetic properties that bear a close resemblance to the electronic properties of semi-conductors, is, naturally, of tremendous scientific interest [9]. Among the most interesting application-oriented prospects of PBG structures, or photonic crystals, are their ability to provide new means of waveguiding. By employment of the PBG effect as compared with the mechanism of total internal reflection (TIR), a completely new class of electromagnetic waveguides has emerged [10-12]. These waveguides possess properties fundamentally different to conventional TIR based waveguides, including the remarkable ability to provide lossless transmission around 90 degree sharp bends [11]. Although not demonstrated at optical wavelengths, such lossless sharply bended waveguides were recently demonstrated experimentally at millimetre wavelengths in planar two-dimensional (2D) photonic crystal structures

[13]. Due to the scalability of photonic crystals this result holds great promise for future realization of highly complex, all optical, integrated circuits.

To fulfill the appealing prospects of high degrees of compactness and integration of optical circuitry using 2D photonic crystal structures, the major problem, at present, concerns achieving a good confinement in the third direction, i.e. in the direction perpendicular to the periodic plane. 2D photonic crystals are, naturally, intrinsically unable to provide this vertical confinement (an essential fact which is nonetheless neglected in most application-oriented 2D photonic crystal investigations). Due to this lack of vertical confinement and the finite height of real 2D photonic crystals, most fabricated PBG structures suffer from high losses [14]. Sandwiching a film of photonic crystal between two low-index material layers is not a sufficient requirement for obtaining leakage-free bound modes in the photonic film. Only under very precise conditions may such bound modes be obtained [15], making careful and detailed three-dimensional (3D) analysis and design of future planar PBG structures vital.

Photonic crystal technology has already emerged as a promising tool for applications in integrated optical technology. The ability to guide light in a manner not possible with traditional waveguides has led scientists to propose the use of this technology in the field of fiber optics. While traditional fiber optics is itself a powerful tool, the use of photonic crystal technology in fibers may lead to a whole range of technology which could never previously been developed. Such technologies would include the remarkable endless single-mode fibers as well as the ability to guide light through a column of air by way of the photonic bandgap properties.

5.8 References

1. For a recent review, see articles in *Photonic Band Gap Materials*, C. M. Soukoulis, ed. (Kluwer, Dordrecht, The Netherlands,1996).

2. J. D. Joannopoulos, R. D. Meade, and J. N. Winn, *Photonic Crystals: Molding the Flow of Light* (Princeton University Press, Princeton, N.J., 1995).

3. M. M. Sigalas, R. Biswas, Q. Li, D. Crouch, W. Leung, R. Jacobs-Woodbury, B. Lough, S. Nielsen, S. McCalmont, G. Tuttle, and K.-M. Ho, "Dipole antennas on photonic bandgap crystals: experiment and simulation," *Microwave Opt.Technol. Lett.* **15**, 153–158 (1997).

4. E. R. Brown and O. B. McMahon, "High zenithal directivity from a dipole antenna on a photonic crystal," *Appl. Phys.Lett.* 68, 1300– 1302 (1994).

5. M.P. Kesler, J.G. Maloney, B.L. Shirley, and G.S. Smith,"Antenna design with the use of photonic bandgap materials as all dielectric planar reflectors,"*Microwave Opt.Technol. Lett.* **11**, 169–174 (1996).Biswas *et al.* Vol. 18, No. 11/November 2001/J. Opt. Soc. Am. B 1689.

6. E Yablonovitch 1987 Inhibited spontaneous emission in solid-state physics and electronics *Phys. Rev. Lett.* **58** 2059–62

7. John S 1987 Strong localization of photons in certain disordered dielectric superlattices *Phys. Rev. Lett.* **58** 2486–9

8. J D Joannopoulos,J N Winn and R D Meade 1995 *Photonic Crystals: Molding the Flow of Light* (Princeton, NJ: Princeton University Press)

9. See e.g. J Rarity and C Weisbuch 1996 *Microcavities and PBGs Physics and applications (NATO ASI Series. Series E, Applied Sciences vol 324)* (Dordrecht: Kluwer)

10. J Broeng, D Mogilevtsev, S E Barkou and A Bjarklev 1999 Photonic crystal fibres: a new class of optical waveguides *Opt. Fiber Technol.* at press

11. A Mekis, J C Chen, I Kurland, S Fan, P R Villeneuve and Joannopoulos J D 1996 High transmission through sharp bends in photonic crystal waveguides *Phys. Rev. Lett.* 77 3787–90

12. H Benisty 1996 Modal analysis of optical guides with two-dimensional photonic band-gap boundaries *J. Appl. Phys.* **79** 7483–92

13. S-Y Lin, E Chow, V Hietala, P R Villeneuve and J D Joannopoulos 1998 Experimental demonstration of guiding and bending of electromagnetic waves in a photonic crystal *Science* **282** 274–6

14. T F Krauss, R M De La Rue and S Brand 1996 Two-dimensional photonic-bandgap structures operating at near-infrared wavelengths *Nature* **383** 699–702

15. P St J Russell, J Pottage, Broeng J Broeng, D Mogilevtsev and P L Philips 1999 Leak-free bound modes in photonic crystal *'99: Conf. on Lasers and Electro-optics (Baltimore, 25–27 May 1999) submitted*

CHAPTER 6

PHOTONIC CRYSTAL RESEARCH

Chapter Content

6. 1. Introduction

A group of scientists is dedicated to working on problems related to a new kind of material, photonic crystals (also known as photonic band-gap materials). Photonic crystals are periodic dielectric structures that have a bandgap that forbids propagation of a certain frequency range of light. This property enables one to control light with amazing facility and produce effects that are impossible with conventional optics.

Photonic crystals are described exactly by Maxwell's Equations, which we can (and do) solve by the application of massive computational power. Much of our research, however, is directed at achieving a higher level of understanding of these systems, so that we can predict and explain their behavior without resorting to brute force calculation. They are also interested in finding new phenomena and devices that are made possible by photonic crystals, and have already filed several patents for their discoveries. They work closely with researchers in the Materials Science and Engineering department of MIT (CMSE) and elsewhere to actually fabricate their designs.

6.2 Research Projects and Results

6.2.1 Resonant Cavities

By making point defects in a photonic crystal, light can be localized, trapped in the defect. The frequency, symmetry, and other properties of the defect mode can be easily tuned to anything desired [1].

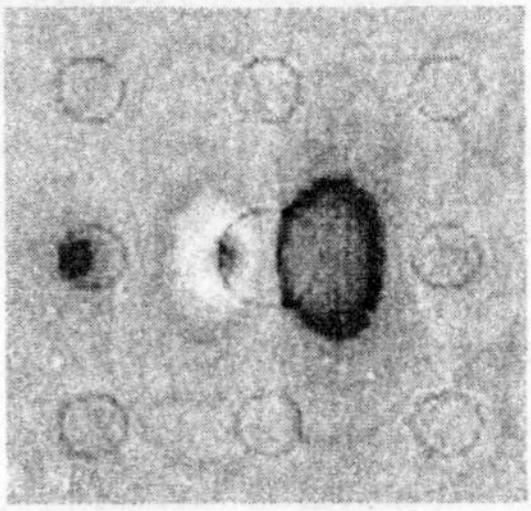

6.2.2 One-dimensionally Periodic Structures

By adding a periodic structure to a conventional waveguide, it is possible to create a one-dimensionally periodic photoniccrystal. Such structures can be used as high-Q filters, and have been successfully fabricated and tested [2].

6.2.3 Three-dimensional Structures

The proposed (in 1994 and 2000) structures with full three-dimensional bandgaps, which will be amenable to fabrication [3] is shown in figure.

6.2.4 Waveguide Bends

With photonic crystals, it is possible to create waveguides that permit 90-degree bends with 100 Per cent transmission. This phenomenon can be understood as the analogue of one-dimensional resonant tunneling phenomena in quantum mechanics [4].

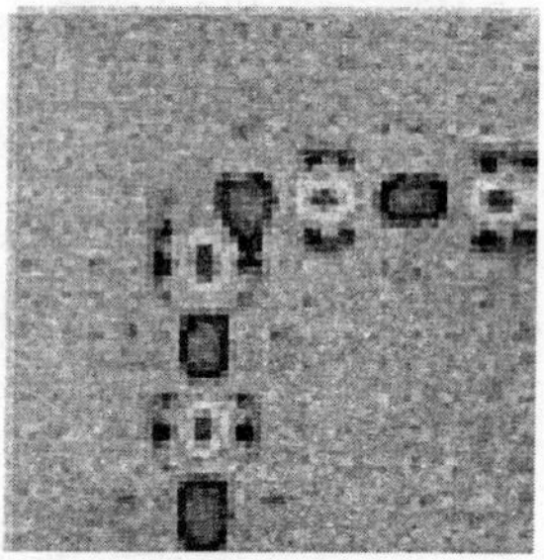

6.2.5 Channel-Drop Filters

Photonic crystals can be used to design a perfect channel-drop filter. This is a device, which picks out a small range of frequencies from a waveguide and reroutes it in another direction, leaving the other frequencies unaffected [5].

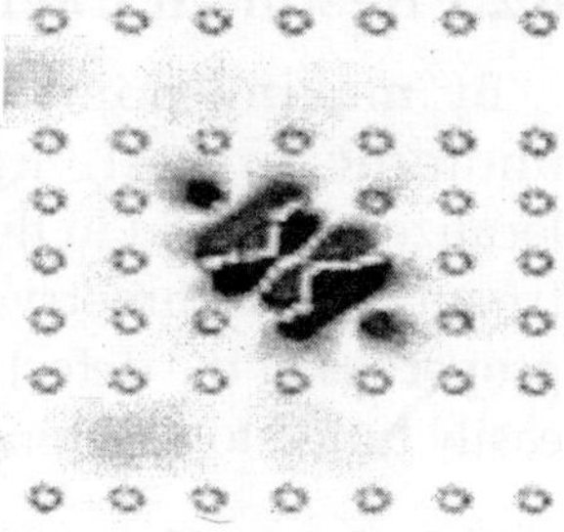

6.2.6 Waveguide Crossings

The proposed novel design for intersecting optical waveguides with negligible crosstalk, using general symmetry considerations that can be applied *a priori* to diverse systems [6] is shown in figure.

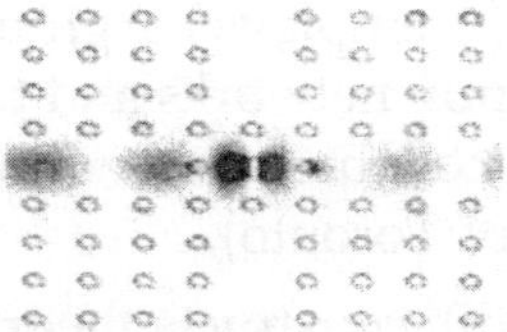

6.2.7 Photonic Micropolis

This is not a research project, but is simply a fanciful depiction of a "photonic micropolis" incorporating many elements of research. Photonic crystal buildings house bundles of light, and highways and bridges guide light along narrowchannels around tight corners [7].

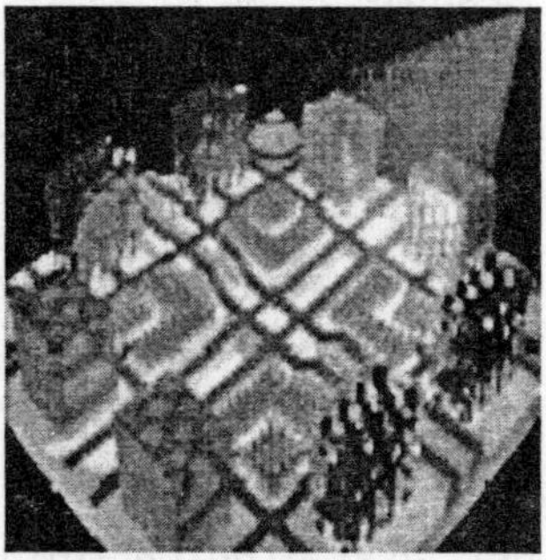

The quantum optics research group in quantum optics has primary interests in quantum informatics and quantum networks, photonic crystals, fundamental issues in quantum physics, open system quantum dynamics, and Lie group techniques. Barry Sanders and Jim Cresser are the two senior members of the group, and the group includes postdoctoral researchers and students. The quantum optics group collaborates with computer scientists at Macquarie University as part of the interdisciplinary Centre for Advanced Computing - Algorithms and Complexity [7].

6.3 Areas of Current Research

6.3.1 Quantum Networks

A quantum network is an array of quantum devices, connected via a structure of communication channels that can be one-way or two-way and can be classical, quantum or a combination of the two. A particularly important example of a quantum network is a distributed quantum computer [8]. Their research (i). employs Lie group theory to characterize complex quantum networks and to analyses their properties and outputs (collaborations with Lakehead University and University of Toronto),

(ii). constructs theories of measurement and detection that are both accurate and applicable to quantum network designs,

(iii). develops and analyses highly non-classical states such as entangled coherent states and their applications to quantum networks,

(iv). considers realizations of quantum networks in both quantum optics and in the area of quantum informatics (with Masaryk University, Stanford University, the University of Queensland and HP Laboratories Bristol), and

(v). investigates the communication structure for a distributed quantum computation in collaboration with the other members of the interdisciplinary Centre for Advanced Computing—Algorithms and Cryptography [8].Some of their key results have included proposals for realizing geometric phase shifts in systems with non-Abelian dynamics, characterizing Bell inequality experiments using operationalism and Lie group theory, formulating quantum computation for qudits and performing detailed analyses of both quantum cryptography, quantum secret sharing and quantum teleportation [8] (which can be key components of the communication structure within the network).

6.3.2 Photonic Crystals

Research into photonic crystals has developed rapidly since the concept was developed in the 1980s, and exciting developments have taken place for photonic crystals in one-, two- and three-dimensional photonic crystal structures. In the most important applications, photonic crystals provide photonic bandgaps, analogous to electronic bandgaps in solid-state physics, and could lead to revolutionary optical circuits [8]. Their research focus is on

(vi). Cavity quantum electrodynamics in photonic crystals and

(vii). Microwave antennae design in photonic crystals

(viii). Their key results include quantitatively accurate theories and modelling of radiating dipoles in photonic crystals [8].

6.3.3 OLED Output Boosted by Photonic Crystal Pattern

Researchers are reporting an increase in the efficiency of an organic light-emitting diode (OLED) by texturing the glass substrate with a photonic-crystal (PC) pattern [9].

6.4 Photonic Crystals - Overview of Processes, Current Research and Potential Industry Applications

Photonic crystals are optical materials with periodic changes in the dielectric constant on a length scale comparable to optical wavelengths. Dimensional types of photonic crystals, parallelism, periodicity, processing methods, nanofabrication, applications and current research are looked at here [10].

1. Ultra Small Quantum Cascade Laser Produced Using Photonic Crystals, Scientists from Bell Labs have built a semiconductor laser that may have numerous applications, from advanced optical communications to sensitive chemical detectors [11].

2. Using Lasers to Make Photonic Crystals From Hydrogel Nanoparticles, A research team from Georgia Tech's School of Chemistry and Biochemistry has created phtonic crystals from hydrogel nanoparticles. By altering the water content of the hydrogels, they found theycouldtune the optical properties of the photonic crystals. [12]

3. **Materials in Information Technology** e.g., Ferroelectric and dielectric ceramics and their applications; Piezoelectric ceramics and transducers Piezoelectric ceramics and transducers; Microwave ceramics; Oxide phosphors; Packaging materials; Magnetic materials; Dielectric ceramics-based photonic crystals; Functional films [13].

4. **Biomaterials and Bio-mimetic**, e.g., Self-assembly of hard-tissue and soft-tissue [13].

5. **Nanomaterials and Nanodevices**, e.g., Nanostructures (e.g., Nanowires); Nano-multilayers; Patterned carbon nanotubes; Polymer-based *Nanocomposites[13].*

6. **Particle-Materials Interaction**, e.g., Ion-beam modification and manipulation of materials [13].

7. **Materials in Energy and Environment**, e.g., Solid electrolyte (polymer-based) materials; SOFC Electrode materials (Li-Co-O, graphites) for batteries; Thermoelectric materials and micro-devices; Catalysts for controlling automotive exhaust gases; Solar cells[13].

8. **Materials in Transportation and Security**, e.g., Colloidal facation processing of low-cost high performance ceramics; Cutting tool of ceramics (e.g., Si3N4); Ceramics for high temperature applications; Glass-ceramics for decorative materials; Functionally graded composites; Nuclear materials; Bainite Steels; Mg alloys; Microcrystalline Al alloys; Ti alloys; Metal-matrix composites [13].

9. **Computation Computational materials science**, e.g., Atomic simulation of solid-state amorphization; Simulation of micromagnetism; Grain boundary and interface; Calculations of effective properties of composites [13].

10. Photonic Crystal Accelerator

At many universities and national labs, electrons are accelerated to high speeds by electric fields imparted by gusts of microwaves. The cavities in which these microwaves are delivered (by devices called klystrons) will support a main radiation mode and other, wakefield, modes, or overtones, as well. Now, however, physicists at MIT have used photonic crystals, material structures which allow the passage of light at some frequencies but not others, to greatly limit overtones in an accelerator cavity. This represents the first time a photonic crystal (also referred to as a photonic bandgap, or PBG) structure has acted as an accelerator. Furthermore, in this case the acceleration gradient, an important measure of an accelerator's efficacy, was 35 MeV/m. This is twice the value one normally obtains at the MIT linac being used for this test, where an electron beam with an energy of 17 MeV was boosted by an additional 1.4 MeV in the photonic-crystal structure, which consists of arrays of tiny rods and operates at a frequency of 17 GHz.

6.5 References

1. http://ab-initio.mit.edu/photons/resonant-cavities.html-Photonic Crystal Research.
2. http://ab-initio.mit.edu/photons/1d-crystal.html-Photonic Crystal Research.
3. http://ab-initio.mit.edu/photons/3d-crystal.html#2000-Photonic Crystal Research.
4. http://ab-initio.mit.edu/photons/bends.html-Photonic Crystal Research.
5. http://ab-initio.mit.edu/photons/ch-drop.html-Photonic Crystal Research.
6. http://ab-initio.mit.edu/photons/cross.html-Photonic Crystal Research.

7. http://ab-initio.mit.edu/photons/micropolis.html- Photonic Crystal Research.

8. http://www.physics.mq.edu.au/~barry/research.html-B.C.Sanders : Research.

9. http://www.azonano.com/details.sap?ArticleID=582-Photonic Crystals 6.

10. http://www.azonano.com/details.sap?ArticleID=1256-Photonic Crystals 6.

11. http://www.azonano.com/details.Sap?ArticleID=62 - Photonic Crystals6.

12. http://www.azonano.com/details.sap?ArticleID=581-Photonic Crystals 6.

13. http://www.mse.tsinghua.deu.cn/en/research/index.htm- Research Areas.

CHAPTER 7

PHOTONIC CRYSTALS IN NATURE

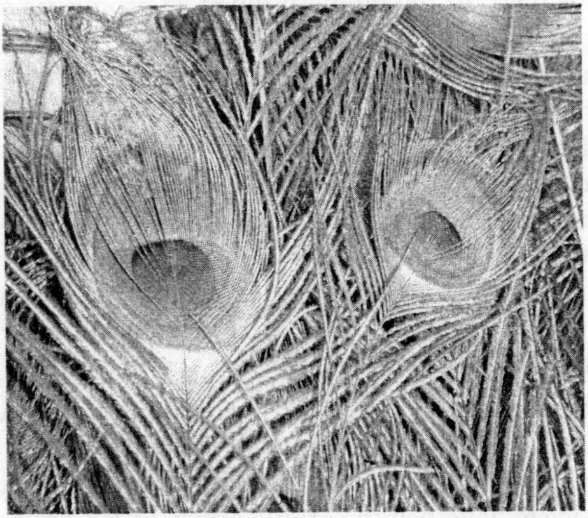

Chapter content

7.1 Introduction

Photonic crystals have been subject of research for the last two decades. These artificial materials behave towards optical electromagnetic radiation in the same way as semi-conductors behave towards electrons. A lot of materials and technologies have been tested for usefulness in this field. At the moment the most promising material is silicon processed with microelectronic technologies. It turned out however, that what people discovered a couple of years ago, Mother Nature created millions of years earlier. Examples of photonic crystals structures occur in hard minerals (opal), as well as in living creatures (butterfly wings, fur of the sea mouse). Naturally created photonic crystal structures aren't all the same, they differ in forms and functions. Two as well as three-dimensional structures are reported. That makes them even more interesting subject for research.

The structural colors of animals have provided an attractive means of advertisement for those reflectors designed and manufactured by humans. Biologists and physicists have begun serious optics-based projects where results will be supplied by evolution. But it is the recent finding of photonic crystals in animals that has really triggered a surge in interest. Animals, it would seem, have plenty to teach us, not only in terms of the design of their optical structures, but also in their engineering. Many organisms have photonic crystals as part of their 'bodies'. Recent research into the diversity of photonic crystals in nature has led us to question how such precise nanostructures form, with the hope that answers may lead to breakthroughs in their engineering.

In a study of structural colors in nature, a remarkable convergence emerged in the nano scale architecture of 2D and 3D periodic photonic crystals between species, families, phyla, and even kingdoms of organisms. Of the many types developed by engineers, living organisms possess only four. These, however, occur in many unrelated species.

Recently even car manufacturers started thing of using natural photonic crystals. In a car show in Sydney, a novel type of paint was used. The vehicle appeared blue from the front...but changed entirely to green when viewed from behind! The manufacturers had replaced pigments with 'structural colors.' The secret behind this coating lies in its self-positioning reflector. Small fragments of thin-films, which cause colored reflections, are mixed within a matrix of transparent paint. When the paint is applied, the thin-films float to the surface so that all are orientated similarly, parallel with the car body, and reflect the same wavelength of light in each direction. As the direction of observation changes, different wavelengths are observed. These thin-film particles are, essentially, butterfly scale analogues.

Animals appear to possess an almost complete anthology of optical devices, as studied by physicists. Now physicists are thinking of incorporating butterfly scales into a paint to provide a uniquely colored coating. Could butterfly scales be 'grown' on an agar plate for this purpose? The four-wheel-drive described above suggests that an application is waiting.

Michael Gale, of CSEM in Zurich, used the example of *Morpho* butterfly scales to highlight the potential of applying nature to commerce as far back as 1989. At the time of his paper, animals were more of a marketing tool than a part of physics. Now things have changed.

Pigments in animals have been well-documented and we are beginning to find that fluorescence in animals has a function after all, making it interesting to biologists and chemists. An animal color is not useful in this sense if it is incidental, i.e. if it has not evolved with the purpose of providing a visual effect (such as the color of blood). But the greatest potential for study by physicists and material scientists lies with animal structural colors. These are made from transparent materials which, ironically, produce the brightest optical effects seen in nature.

7.2 Butterfly Crystals

Birds do it, beetles do it, maybe not educated fleas, but certainly butterflies do it. That is, they use clever structural engineering in their body parts to produce iridescent colors without all the fuss of synthesizing complex pigment molecules. Physicists refer to the microscopic features on the surface of a butterfly's wing, say, as photonic crystals.

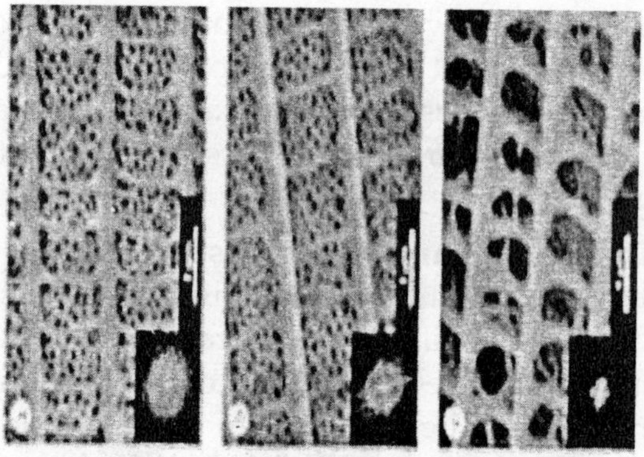

Figure 7.1: Wings of Butterfly under SEM

Unlike pigments, which absorb or reflect certain frequencies of light as a result of their chemical composition, photonic crystals reflect light because of their surface characteristics. The crystal array may reflect more blue light from one angle than another so the fluttering of a butterfly or the dash of a kingfisher will produce a beautiful array of colors. Physicists would like to be able to emulate natural photonic crystals for their own purposes but could also help biologists better understand how different species make use of iridescence.

The researchers used high-resolution scanning electron microscopy to take a close look at the wings of the butterfly popularly known as Meleager's Blue (*Meleageria daphnis;* but properly classified as *Polyommatus daphnis*) and confirmed that the surface of the wings are coated with tiny scales pitted with arrays of submicrometre-sized holes, which form the structure of the natural photonic crystal, and give the male wing a shimmery blue sheen.

The wings of the closely related butterfly from higher altitudes (2000-2500m), the Persian Polyommatus marcidus, however, lack these holes, and its wings are a dull brown. "The blue colour can be attributed unambiguously to the fine, sponge-like medium, called the "pepper-pot structure", present between the ridges and the cross ribs in the scales of the coloured butterfly(Figure 7.2). Only traces of this structure can be found on the scales of the discoloured butterfly," explain the researchers.

Figure 7.2: The blue coloured butterfly

Photonic crystal type nano architectures (built of chitin and air) occurring in the scales of two male butterflies, with dorsal blue and ventral green coloration were investigated by optical microscopy, SEM and TEM. Reflectance data were collected by normal incidence measurements and by integrated reflectance measurements, too. Reflectance peaks with close positions in the blue range of the spectrum were found both for the single crystalline scales and for the most disordered photonic band gap material. Polycrystalline structures with grain size on the micron scale produce matt reflectors, while moderately ordered "pepper-pot" type structures can yield surfaces with shiny aspect. It was found that the cover scales on each sides of the wing are primarily responsible for the observed colors. The character of the observed colors (shiny or matt) correlates with the habitat of the investigated butterflies.

A single iridescent butterfly scale has longitudinal ribs jutting out from the base of the scale perpendicularly and running parallel with each other. These ribs bear side branches, forming the high index layers of a quarter-wave stack (air forms the low index layers). For decades this explanation represented the end of story, until recently. Roy Sambles and Peter Vukusic of Exeter University, along with Chris Lawrence of QinetiQ, have found a way to measure the reflectance from a single butterfly scale. Rather than use an ordinary light beam, where the beam waist cannot be reduced to less than a scale width, they employ laser beams of different wavelengths. Along its path, the laser beam is deflected up, forward, then back down to create space for a detector to move 360° around the scale. The angular reflectance measurements are accurate and cannot always be explained by mirror reflection laws. There must be more to some butterfly scales than quarter wave stacks, and the body of the scale beneath the ribs appears to hold the answer.

The explanation lies in adaptation of each species to its environment. At the warmer lower altitudes, the butterfly can afford to be showy and, adds biologist Zsolt Balint of the Hungarian Natural History Museum, "Even with their high reflectance they can control overheating." But, its cousins eking out a living up the mountain can get warmer in the sun by spreading their dull brown wings to absorb previous heat. The researchers suggest that the butterflies at high elevations trade flashy iridescence for light-absorbing brown so that they can withstand colder temperatures and garner enough energy during short sunny periods to survive long enough to mate.

This temperature effect might be put to good use in novel clothing materials that have a photonic crystal coat allowing the wearer to stay cooler in the desert by reflecting heat-inducing light, but be switched for a non-iridescent form to help absorb heat in the cold of space, for instance.

7.3 Why are Some Coleoptera Beetles Blue?

Because light striking the beetle's external hard parts undergoes destructive interference. Precisely how this happens is now being studied quantitatively by a team of scientists in Namur, Belgium. Electron microscope pictures of the beetle's scaly cuticle help to explain that each scale is made of alternating layers of pure chitin (high index of refraction) and mixtures of chitin and air (low index of refraction).

The resulting structure is a photonic crystal: because of wave interference, light of certain frequencies are excluded. In this case blue light is forbidden from being absorbed by the animal's shell; all blue light is reflected while other frequencies are absorbed in the cuticle, and the creature consequently has a blue appearance.

Lord Rayleigh in 1918 was the first to suggest that the iridescence of some insects arose from interference effects. And by now the photonic-crystal effect is known to occur naturally in many places, such as the opalescence of weevils and the striking colors in the peacock's tail feathers (Figure 7.3).

Figure 7.3: The Peacock

7.4 Animal optics

Although Hooke and Newton initiated the field in the Seventeeth century, it was not until the introduction of the electron microscope to biology in the 1940s that structural colors in animals became a subject of study as such. Prior to that, however, eyes had attracted significant attention, with the elucidation of compound eyes of some insects and the graded-refractive-index lenses of fish. Biologists have also found some variations on insect eyes, not to mention a telephoto lens component in a jumping spider eye. The most recent discovery is the micro-lens array of the bristle star— the closely packed, tiny convex lenses that are used as magnifiers in human applications.

Animals' metallic-like look of structural colors must be due to refractive index differences, diffraction, and interference. The most common and well-known form of animal reflector is the multilayer type. This involves stacks of thin-films, with thicknesses around a quarter of the wavelength of light and alternating in high and low index materials. The difference in refractive index in the stack is just 0.2–0.5, making animals interesting to physicists from this point alone. In fact, this low contrast is common to all types of animal reflectors.

Animal multilayer reflectors include the quarter-wave stack, where a spectrum is dispersed, and broadband or silver/gold reflectors (mirrors). Broadband reflectors incorporate layers of various thicknesses, ranging from thin-films for blue to those of red, which can be stacked randomly or systematically. In some cases, individual layers may even move around within the broadband stack, offering flexibility to the reflector.

Multilayer reflectors are under design and construction for use as mirrors, filters, polarizers, security items, and decorative films. But the right hand of science can be unaware of its left.

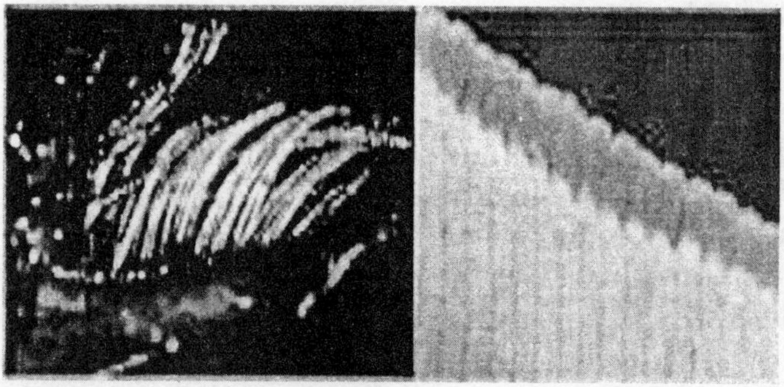

Figure 7.4. The iridescent effect from the first antenna of the seed-shrimp Azygocypridina lowryi is shown on the left. On the right, a scanning electron micrograph shows the diffraction grating on a single hair, with a ridge spacing of 600nm.

Some relatively simple optical reflectors have gradually increased in efficiency over millions of years of evolutionary fine-tuning. For example, the corrugations on the surface of seed-shrimp (cypridinid ostracod) hairs form a diffraction grating and cause iridescence (see Figure 7.4), which is used as a courtship display. These gratings have improved in optical efficiency throughout the evolution of the group. Conceivably, this is the result of minor changes in the genome, with each change trans-lating to a minor change in the structure. Other simple optical reflectors could be considered to evolve in a similar, gradual manner, where they are manufactured via periodic deposition of materials.

Since their discovery in seed-shrimps, diffraction gratings have emerged in a diversity of animals from worms to flies, often occurring on hairs. Again nature has thrown up some unusual designs, unknown to physics. The low contrast in refractive indices involved offers considerable hope for reproducing these animal structures in the laboratory and on a commercial scale. The precise materials of the animals, often involving chitin, do not require reproduction, and if the refractive index gap can be widened by substituting plastics or glasses, then optical effects may be enhanced.

It is possible that photonic nanostructures can evolve without passing through the string of intermediate stages one might expect from a complex architecture. It is the fact that nature's photonic crystals are made by individual cells that is so appealing to the engineer. If we can keep these cells alive in vitro, and induce them to reproduce in some cases, conceivably they will make photonic crystals to order.

Science fiction is becoming science, and the next results to emerge from this subject could make it a hot topic.

7.5 References

1. A. R. Parker, Discovery of functional iridescence and its coevolution with eyes in the phytogeny of Ostracoda (Crustacea), *Proc. R. Soc. Lond.* B 262, pp. 349-355, 1955.
2. Parker, A.R. Parker, "The Cambrian light switch." *Biologist* (1999) **46** (1), 26-30

3. A.R. Parker, "Light-reflection strategies." *American Scientist* (1999) 87 (3), 248-255

4. Parker, A.R., "515 million years of structural color." *J. Optics A: Pure and Applied Optics* (2000) **2**, R15-28

5. M.F. Land, and D.-E. Nilsson, (2001) *Animal Eyes* (Oxford University Press) Oxford

6. A.R. Parker, (2003) *In the Blink of an Eye* (Simon & Schuster).

7. L. T. Threadgold, The ultrastructure of the animal cell, Pregamon press, Oxford, 1967.

8. P. L. A. M. Corstjens and E. L. Gonzales, Effects of nitrogen phosphorus avail ability on the expression of the -vesicle V-ATPase (subunit c) of Pleurochrysis (Haptophyta) 1, *J. Phycology* 40, pp. 82-87, 2004.

9. G. Gemne, Ontogenesis of corneal surface ultrastructure in Lepidoptera, *Phil. Trans. R. Soc. Lond.* B 262, pp. 343-363,1971.